WHALES AND DOLPHINS OF THE EUROPEAN ATLANTIC

The Bay of Biscay, English Channel, Celtic Sea and coastal SW Ireland

Dylan Walker & Graeme Cresswell

WILDGuides

First published 2001 by **WILD**_Guides_ Ltd.
Second Edition 2008
Reprinted 2011

WILD_Guides_ Ltd.
Parr House
63 Hatch Lane
Old Basing
Hampshire
RG24 7EB
UK

www.wildguides.co.uk

ISBN 978-1-903657-31-7

© 2008 Dylan Walker and Graeme Cresswell
Robert Still (illustrations, maps and digital artwork)
Designed by Robert Still
Edited by Andy Swash

Copyright in the photographs remains with the individual photographers.

A catalogue record for this book is available from the British Library.

All rights reserved. No part of this publication may be reproduced, stored in a retrieval system, or transmitted, in any form by any means, electronic, mechanical, photocopying, recording, or otherwise, without the prior permission of the publishers.

Printed and bound by ABC Print, England.

Contents

About this guide 4
How to use this guide 5
The European Atlantic 6
What is a cetacean? 8

WHALE AND DOLPHIN WATCHING
Where to look 9
When to look 10
Weather 11
What to look for: identification 12
What to look for: behaviour 14
How to look 15
Cetacean families 18

REGULARLY OCCURRING SPECIES
The Great Whales 22
MINKE WHALE *Balaenoptera acutorostrata* 24
SEI WHALE *Balaenoptera borealis* 26
FIN WHALE *Balaenoptera physalus* 28
HUMPBACK WHALE *Megaptera novaeangliae* 30
SPERM WHALE *Physeter macrocephalus* 32
The Beaked Whales 34
The Blackfish 35
NORTHERN BOTTLENOSE WHALE *Hyperoodon ampullatus* ... 36
CUVIER'S BEAKED WHALE *Ziphius cavirostris* 38
Identifying *Mesoplodon* Beaked Whales 40
SOWERBY'S BEAKED WHALE *Mesoplodon bidens* 42
TRUE'S BEAKED WHALE *Mesoplodon mirus* 43
KILLER WHALE or ORCA *Orcinus orca* 44
LONG-FINNED PILOT WHALE *Globicephala melas* 46

Dolphins, the *Kogia* sperm whales and Beluga 48
SHORT-BEAKED COMMON DOLPHIN *Delphinus delphis* 50
STRIPED DOLPHIN *Stenella coeruleoalba* 52
BOTTLENOSE DOLPHIN *Tursiops truncatus* 54
RISSO'S DOLPHIN *Grampus griseus* 56
ATLANTIC WHITE-SIDED DOLPHIN *Lagenorhynchus acutus* .. 58
WHITE-BEAKED DOLPHIN *Lagenorhynchus albirostris* 60
HARBOUR PORPOISE *Phocoena phocoena* 62

RARE SPECIES
BELUGA *Delphinapteras leucas* 65
BLUE WHALE *Balaenoptera musculus* 66
NORTH ATLANTIC RIGHT WHALE *Eubalaena glacialis* 68
PYGMY SPERM WHALE *Kogia breviceps* 70
DWARF SPERM WHALE *Kogia simus* 71
GERVAIS' BEAKED WHALE *Mesoplodon europaeus* 72
BLAINVILLE'S BEAKED WHALE *Mesoplodon densirostris* 73
SHORT-FINNED PILOT WHALE *Globicephala macrorhynchus* . 74
FALSE KILLER WHALE *Pseudorca crassidens* 75
PYGMY KILLER WHALE *Feresa attenuata* 76
MELON-HEADED WHALE *Peponocephala electra* 77
ROUGH-TOOTHED DOLPHIN *Steno bredanensis* 78
FRASER'S DOLPHIN *Lagenodelphis hosei* 79

Glossary 80
Cetacean conservation – and how you can get involved ... 81
Further reading 82
Useful websites 83
Photographic and artwork credits 84
Index 86
Acknowledgements 88

About this Guide

This is the first comprehensive guide to the identification of whales, dolphins and porpoises (collectively known as cetaceans) in the European Atlantic. Until very recently, most researchers and whale-watchers were unaware of the great variety of cetaceans that can be seen so close to the shores of western Europe. Indeed, it is only during the last decade, when detailed cetacean surveys have been carried out in earnest, that we have discovered how important this area is for cetacean biodiversity.

This field guide describes all of the 31 species of whale, dolphin and porpoise that have occurred in the European Atlantic. Knowledge of the identification of cetaceans at sea is developing rapidly and this book aims to provide an easy-to-use, detailed guide for both the ocean-going and land-based whale-watcher.

Because whales and dolphins often only show a small proportion of their bodies at the surface, and may only be seen briefly, observers have in the past often considered definite identification to be too difficult. Although some species are rarely seen at sea, either because they avoid boats, live far from shore or remain submerged for long periods of time, and their identification is still poorly known, in most cases identifying cetaceans at sea is not always as difficult as it may first appear.

There are many features to look for and, with experience, it is often easy to identify a whale or dolphin at a considerable distance. Although little is known of the biology and identification of some species within the European Atlantic, our understanding of these mysterious animals is increasing all the time.

It is hoped that this guide will encourage more whale-watchers to explore the waters of the European Atlantic, thereby adding to our knowledge of the status and distribution of these wonderful animals.

If you have good quality photographs of any of the species described in this book, or of other cetaceans from further afield which might be suitable for inclusion in future publications, please email the authors: photos@wildguides.co.uk.

For example, this is the first publication ever to show photographs taken at sea of the four species of *Mesoplodon* beaked whale currently known to occur in the North Atlantic. These photographs have provided some new pointers to the identification of this particularly difficult group.

How to use this guide

Following sections covering whale and dolphin-watching and cetacean behaviour (*pages 9–17*) and an introduction to the cetacean families (*pages 18–21*) the 18 regularly occurring species (*pages 22–63*) are covered as shown opposite. Rare species (*pages 64–78*) are covered in a similar way.

Text covers the following:

BEHAVIOUR: A description of the typical activities of the species.

STATUS AND DISTRIBUTION: A summary of seasonal distribution, habitat preferences and any migration patterns together with relevant notes on ecology and conservation.

IDENTIFICATION: A summary of each species key identification features and the most noticeable characteristics accompany the photographs.

SIMILAR SPECIES: A detailed summary of the differences to look for between confusing species to enable a confident identification.

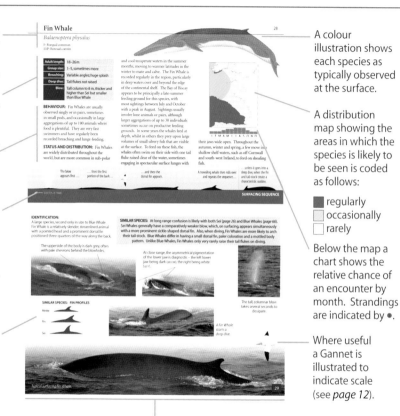

For every species the English name, scientific name, French name and Spanish name are given.

Key information on length, group size, diving behaviour and blow is shown where appropriate.

Where required a species' surfacing sequence is shown by annotated silhouettes with a Gannet to indicate scale.

Small illustrations to enable comparison between similar species are provided where useful.

Facing each species account is a colour plate which has been created, using carefully selected photographs, to illustrate key identification features.

Although photographs of cetaceans at sea usually only reveal a part of the raised body or merely a blow, such images are considered to be the most informative way of aiding identification, since they show the animal from the whale-watcher's perspective.

A colour illustration shows each species as typically observed at the surface.

A distribution map showing the areas in which the species is likely to be seen is coded as follows:

- regularly
- occasionally
- rarely

Below the map a chart shows the relative chance of an encounter by month. Strandings are indicated by ●.

Where useful a Gannet is illustrated to indicate scale (see *page 12*).

The European Atlantic

For the purposes of this book, the European Atlantic is divided into four regions based on their differing topography.

Each region differs in habitat characteristics (see map opposite and text below).

English Channel and the seas west to Ireland

The Atlantic coasts of southern Ireland, Wales and southern England face out towards the Celtic Sea, Irish Sea and the English Channel respectively. These seas form an area of shallow water over the European continental shelf at a depth generally less than 100 metres.

Northern Bay of Biscay

The Northern Bay lies between the 200 metre depth contour in the north and 44°30'N in the south. The steep topography of the shelf-edge here results in a sharp increase in water depth from 200 metres to 4,000 metres, before levelling out. Beyond this lies the abyssal plain of the ocean floor.

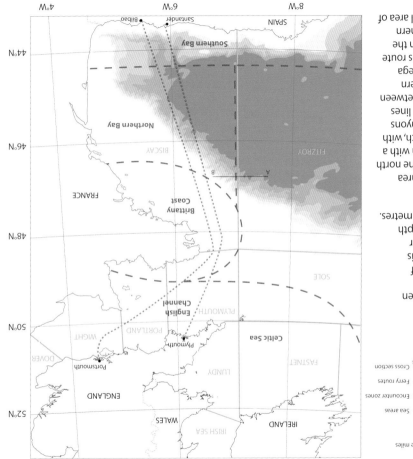

Brittany Coast of France

The Brittany Coast lies between 48°40'N and the 200 metre depth contour at a latitude of approximately 46°20'N. This is also a shallow water area over the continental shelf, with depth averaging between 100–150 metres.

Southern Bay of Biscay

The Southern Bay covers an area from 44°30'N, southward to the north coast of Spain. This is an area with a tremendous variation in depth, with two extensive deepwater canyons that fall beneath the transect lines of the two ferries which ply between southern England and northern Spain. These are the Torrelavega Canyon on the Brittany Ferries route and the Santander Canyon on the P&O Ferries route. The Southern Bay also encompasses a small area of continental shelf.

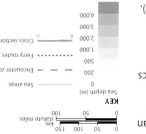

When discussing the distribution of cetacean species, these regions will be referred to regularly (see map opposite and text below).

CROSS SECTION SHOWING THE TOPOGRAPHY OF THE NORTHERN BAY OF BISCAY

Abyssal plain

This is the ocean floor beyond the continental shelf. These waters are the deepest of all, and are generally over 1,000 metres deep.

Prey species such as migratory fish and squid occur unpredictably and patchily in these waters.

Cetaceans include deep-diving squid-eating whales, migratory baleen whales, and nomadic herds of dolphins, all of which are capable of searching large areas for their prey.

Continental slope

Also known as the shelf edge or shelf slope, this is a region of incline between the continental shelf and the abyssal plain.

Like a giant cliff or ridge of underwater mountains, the continental slope separates the continental shelf from the abyssal plain. Amongst the series of ridges and valleys can be found steep-sided canyons, inhabited by a high density of squid and favoured by squid-eating species such as beaked and sperm whales.

The continental slope forces currents to rise from the ocean floor in a process called upwelling. These currents may carry nutrients that are seized upon by phytoplankton concentrated near the surface where sunlight penetrates and enables photosynthesis to take place.

The resulting food chain includes whales, dolphins and seabirds, which often concentrate in these areas in great numbers.

Continental shelf

This is a horizontal ledge under shallow water between a continental landmass and the ocean floor. All coastlines within the European Atlantic overlook shallow seas on the continental shelf.

These waters are relatively productive for cetacean prey due to the availability of nutrients from sediments on the sea floor and from major rivers. These sediments become suspended in the water column where currents are strong or at times of year when the sea becomes rougher. Subsequently they provide sustenance for the marine food chain.

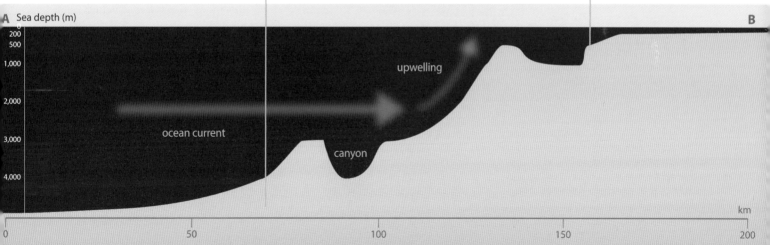

What is a cetacean?

All whales, dolphins and porpoises belong to an Order of marine mammals called Cetacea, and are collectively known as cetaceans. Like other mammals, cetaceans are warm-blooded, breathe air through their lungs and suckle their young. However, unlike most other mammals, cetaceans live exclusively in water.

Baleen whales range in size from the Pygmy Right Whale, which barely reaches 7 m in length, to the massive Blue Whale, which grows to over 30 m. Instead of teeth, these whales have plates of baleen (whalebone) which hang from the roof of their mouths. These vertical plates can grow to over 2 m in length in some species and are used to filter enormous quantities of small fish and crustaceans. Baleen whales have two external nostrils or blowholes.

Toothed whales range in size from the huge Sperm Whale, with a maximum length of 18 m, to the diminutive Harbour Porpoise, which reaches a maximum length of 1.8 m. Besides having teeth, the Odontoceti are distinguished by having only one external nostril or blowhole. Nearly 70 species of toothed whale are recognised, the smallest of which are generally known as dolphins or porpoises.

There are approximately 80 species within the Order Cetacea. These are divided into two groups or Sub-orders: the Mysticeti or baleen whales, and the Odontoceti or toothed whales.

Their adaptations for life in the oceans are so advanced that many cetaceans superficially resemble fish, particularly sharks, in form and structure. One of the best ways of differentiating a cetacean from a fish is by looking at the orientation of the tail: horizontal in cetaceans, vertical in fish.

The following diagrams illustrate the main differences between baleen and toothed whales, and give the names of the key external body parts which can be important in identification.

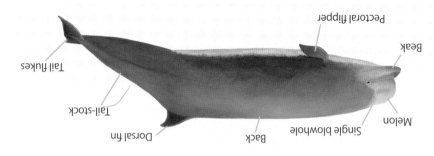

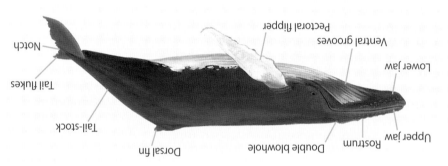

Whale and Dolphin Watching: where to look

Where to look

Most cetaceans, like their mammalian relatives on land, are at the top of the food chain. Such predators tend to be sparsely distributed and difficult to find, so knowing where to look is very important.

Some cetaceans favour shallow waters or coastal areas, whereas others prefer the shelf slope, or restrict themselves to very deep water. Some are resident, while others are migratory. With experience, and the aid of marine charts and global positioning systems, the whale-watcher will become increasingly familiar with habitat preferences which can help with identification. For example, Minke and Northern Bottlenose Whales look almost identical when only the upper back and dorsal fin is seen, but Minke Whales generally occur over the shelf, often close to shore, whereas Northern Bottlenose Whales are more likely to be encountered in deep offshore waters.

Although water depth and sea floor topography can be useful indicators of the presence of a particular species, this alone should never be used as a way of identifying a cetacean specifically. Part of the joy of watching whales and dolphins is their complete unpredictability. It is almost impossible to guess where they will be because they can pop up anywhere at any time. Some European dolphins have even been recorded over 100 kilometres up river!

Whales and dolphins do not follow any rules. They roam freely over great distances and there is still much to learn about their distribution and migratory patterns. For the dedicated whale-watcher, the most important principle is to put in as much effort as possible. The longer you watch, the better your chances are of seeing something. It is as simple as that!

When observing from land it can be particularly useful to note the state of the tide in relation to cetacean activity. Three species that regularly occur close to shore – Minke Whale, Bottlenose Dolphin, and Harbour Porpoise – all take advantage of fast-moving currents to hunt their prey.

Cetaceans can occur close to land, such as these Killer Whales – putting in the hours may bring good fortune!

Whale and Dolphin Watching: when to look

When to look

Whales, dolphins and porpoises must return to the surface to breathe on a regular basis, so can be seen at any time of day. The amount of time spent at the surface varies greatly from species to species and is also dependent upon whether the animal is resting, feeding or travelling.

However, perhaps the most important factor in terms of when to watch whales is the season. Different species occur in different waters in different months, so the dedicated whale-watcher is likely to encounter a greater diversity of cetaceans by watching in all seasons. It is widely accepted that in temperate regions of the world, such as the European Atlantic, summer, and particularly late summer, is the best time for whale-watching.

There are several reasons for this: the weather, and therefore the sea, is generally at its calmest; high-pressure weather systems often settle for considerable periods bringing warm, sunny and windless conditions; and day length is at its greatest – allowing more hours for whale-watching during a voyage. These factors combine to increase water temperature as summer progresses, allowing phytoplankton (the 'green grass' of the marine ecosystem) to flourish. At this time of year, these free-floating marine plants increase in numbers dramatically, providing food, either directly or indirectly, for all marine creatures ranging from microscopic zooplankton to 100 tonne whales.

This great abundance of food is often concentrated in certain areas, particularly where ocean currents rich in nutrients are forced to the surface. It is in such areas that whales and dolphins often congregate and it is therefore unsurprising that the peak months for whale-watching in the European Atlantic are from June to October.

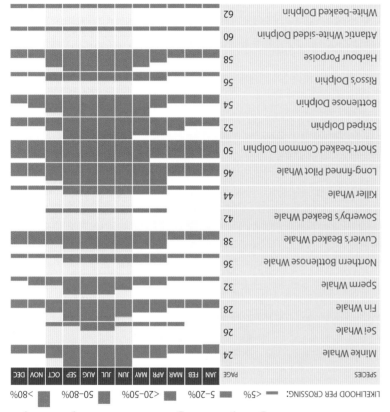

Encounter rate of regularly occurring cetaceans in the Bay of Biscay

LIKELIHOOD PER CROSSING: <5% 5-20% <20-50% 50-80% <80%

SPECIES	PAGE	JAN	FEB	MAR	APR	MAY	JUN	JUL	AUG	SEP	OCT	NOV	DEC
Minke Whale	24												
Sei Whale	26												
Fin Whale	28												
Sperm Whale	32												
Northern Bottlenose Whale	36												
Cuvier's Beaked Whale	38												
Sowerby's Beaked Whale	42												
Killer Whale	44												
Long-finned Pilot Whale	46												
Short-beaked Common Dolphin	50												
Striped Dolphin	52												
Bottlenose Dolphin	54												
Risso's Dolphin	56												
Harbour Porpoise	58												
Atlantic White-sided Dolphin	60												
White-beaked Dolphin	62												

Whale and Dolphin Watching: weather

Weather

Weather conditions in the European Atlantic vary considerably through the year. Winter generally sees a series of wet and windy fronts pushing north-east across the North Atlantic, creating choppy to stormy conditions for long periods. As spring approaches, the weather improves and by mid-summer sunny days and calm seas predominate.

In July, August and early September these conditions can last for weeks, and are perfect for whale-watching. From mid-September the weather starts to become more unpredictable, worsening slightly as autumn turns to winter. The Sun's glare, fog, sea swell, rain and wind can all affect the chances of locating whales and dolphins, but by far the most important factor is the sea state.

Sea state is the term used to describe the wave formation created by the wind. A sea state of three or less, when the waves have few or no white caps, is generally considered to be most conducive to whale-watching and cetacean surveying. Anything higher and it is certainly easy to miss at least the diminutive Harbour Porpoise.

Sea state can be measured using the following guide:

0. Mirror calm; whale-watching heaven!
1. Slight ripples; no white water
2. Small wavelets; glassy crests, no white caps
3. Large wavelets; crests begin to break; few white caps
4. Longer waves; many white caps; whale-watching becomes more tricky
5. Moderate waves of longer form; some spray
6. Large waves; many white caps; frequent spray
7. Sea heaps up; white foam blows in streak
8. Long, high waves; edges breaking; foam blows in streaks
9. High waves; sea begins to roll; dense foam streaks; scary!

What to look for: identification

Sightings of cetaceans are often brief, and generally involve seeing only a proportion of the animal at any one time. It is therefore important to concentrate on several key features in order to reach a conclusive identification. In approximate order of importance, these are:

Size

Due to the relatively uniform nature of the surface of the sea, judging the size of a cetacean can be extremely difficult. Try to compare the size of the cetacean with other familiar objects at a similar distance, such as fishing boats or seabirds.

When observing a cetacean, remember that often only a small portion of the animal will be visible above the surface at any one time. For example, a Blue Whale rolling at the surface will often show less of its body than does the smaller Fin Whale, giving a misleading impression that the former is the smaller animal.

Comparing a cetacean with an object of known size such as a boat or seabird can help in the estimation of size. In the photo above a boatload of people enjoy a simultaneous encounter with a White-beaked Dolphin and a Humpback Whale.

In the European Atlantic there is plenty of opportunity at sea to observe Gannets which have a 'handy' 2 metre wingspan (very useful when judging size). The illustration of an adult is superimposed to scale on the photo. The same illustration is used throughout the book where showing scale or size is important.

Shape

Cetaceans come in many shapes, ranging from long and slim, through rotund and bulky, to short and sleek. It is, therefore, very important to try to judge the shape of an animal.

A Minke Whale (LEFT) appears broader across the back than the similar sized Cuvier's Beaked Whale (RIGHT).

A process of elimination

Correctly identifying cetaceans at sea requires patience and experience. Even experts regularly fail to see sufficient features to identify an animal, and observers will often try to put a name to a sighting based on just one feature will often be incorrect in their assumption.

This is because making judgements at sea is difficult, and also because cetaceans have a cheeky habit of breaking the identification rules that we apply to them. For example:
• Short-beaked Common Dolphins typically have a criss-cross pattern on their flanks, but all kinds of weird patterns have been seen;
• Fin Whales are known not to raise their tail flukes clear of the water during deep dives, but on rare occasions they do;
• Dolphins may be the bow-riding specialists but the Harbour Porpoise, that normally shies away from boats, has been occasionally observed riding the pressure waves of vessels.

These challenges should not put off the beginner, as they are what make identifying a relatively small number of cetacean species so much fun!

No matter what your experience, there is always room to learn something new, and, as long as you follow the simple rule of noting down a number of features in order to reach a positive identification, you will be able to avoid many of the pitfalls.

Coloration and patterning

The colour of a cetacean can be very difficult to judge. Since the sea often looks very different on the port (left) and starboard (right) sides of a ship due to the position of the sun, an observer may be reduced to using the terms light and dark to describe a cetacean. However, coloration and patterning are key features in the identification of some cetaceans, particularly members of the dolphin family which can otherwise appear very similar.

Dorsal fin

The size, shape and position of the dorsal fin is a very useful aid to identification, particularly in whales. Apart from both the extremely rare North Atlantic Right Whale and Beluga, all cetacean species occurring in the European Atlantic show a dorsal fin. The shape of this fin varies greatly and can be triangular, sickle-shaped or merely a hump. In some species, there is often a considerable variation in fin shape and size between individuals. Try to determine the size of the fin in relation to the length of the body, its position along the back and, in large whales, its appearance at the surface in relation to the blowhole.

Two Fin Whales: a typical 'columnar' blow (TOP) and a wind-affected 'bushy' blow (BOTTOM).

Bottlenose Dolphins are grey (ABOVE), but in evening light they can appear brown (BELOW).

Rorqual (MINKE WHALE)

Beaked Whale (CUVIER'S)

Blackfish (SHORT-FINNED PILOT WHALE)

Dolphin (SHORT-BEAKED COMMON)

Additional clues

There are a number of additional clues that indicate that cetaceans are more likely to be present. The most important of these are the presence of other predators of fish, such as circling flocks of seabirds, sightings of seals, or fishing boats. At times the fish themselves may be observed leaping at the surface.

Blow

In many of the large cetaceans, a tall jet of water is expelled as the animal surfaces and breathes out. This is known as the blow and can be a helpful clue to identification in combination with other features. Blow size and shape varies between species but it is also affected by other factors: an animal surfacing after a deep dive will usually exhibit a larger blow than a resting individual; younger animals are smaller and therefore on average have a smaller blow; wind can modify the shape, size and angle of the blow. Because blows are so variable and may be affected by the wind, they should never be used in isolation to identify a whale to species.

What to look for: behaviour

Most cetaceans are highly social animals and exhibit a variety of interesting and distinct behaviours, some of which are depicted below. Cetacean behaviour can sometimes be used to aid identification. For example, some large whales such as Blue, Humpback and Sperm Whales, regularly raise their tails on diving (fluking), unlike Fin, Sei and Minke Whales.

Bow-riding: Swimming in the pressure wave created ahead of large objects pushing through the water such as ships and whales. This behaviour, a speciality of most dolphins, is useful in assisting with identification as some species and populations are keener to ride the bow than others.

Blowing or spouting (see *page 13*): The breath of a cetacean, in which moisture-laden air is expelled from the lungs as a visible spout of water. The blow of the larger whales can often be seen at several kilometres distance.

Breaching: The act of propelling the body upwards until most or all of it is clear of the water.

Chorus line: The co-ordinated movement of a group of cetaceans as they surface in a line-abreast. This behaviour is frequently observed in members of the blackfish family and the Risso's Dolphin, often when they are searching for prey.

Fluking: Some species of whale regularly raise their tail flukes vertically into the air as they dive. The shape and colour of the flukes and the patterning on the undersides are useful identification features.

Lobtailing: The act of a large whale lifting its tail high out of the water before slapping

the flukes against the surface. This is often repeated many times.

Logging or rafting: Resting motionless at the water's surface in a horizontal position.

Lunge-feeding: The explosive movement of a whale as it rises clear of the water with its jaws wide open to catch large schools of fish or invertebrates. Lunge-feeding is particularly characteristic of the rorqual whales. During a lunge they are capable of expanding their throats like a balloon as the whale takes in a large volume of prey and seawater simultaneously, before seawater is sifted out of the mouth through the baleen plates.

Porpoising: Very fast movement involving arc-shaped leaps clear of the water with a clean, head-first re-entry. This behaviour is most commonly seen in dolphins.

Spy-hopping: Raising the head vertically out of the water high enough for the eyes to view above the surface. The head usually then sinks below without making a splash.

Tail-slapping: Small cetaceans, particularly dolphins, are capable of lifting their tail flukes above the water and bringing them crashing down. This behaviour may be repeated many times in a single session.

Some examples of behaviour (TOP TO BOTTOM) *Breaching Humpback Whale; Spy-hopping Long-finned Pilot Whale; Tail-slapping Long-finned Pilot Whale.*

Whale and Dolphin Watching: how to look

How

When searching for cetaceans, it is very important to choose a good place from which to watch. On land, the best locations are headlands with a reasonably high vantage point. Binoculars and often a telescope are useful tools for land-based watching. If you are watching from a boat, you should always carry binoculars. The bridge or close to the bow are good places from which to search, since this allows the observer to look forward and beyond the bow. A stable platform helps the observer to use binoculars without too much shake. Elevation is also an important consideration, particularly in rougher seas when a high vantage-point allows the observer to track cetaceans as they travel through wave troughs. Other important factors include sheltering from the wind, and avoiding glare from the sun and sea spray.

Finding cetaceans can require a great deal of patience. Keep searching over the same area of water even if you think there is nothing there. Some cetaceans can stay under water for an hour or more before eventually surfacing, and even when animals remain at the surface it is surprising how easy they are to miss. A 15 m long Sperm Whale can easily be overlooked when it lies motionless at the surface with only a small portion of its upper back showing.

Cetacean watching is generally easiest in relatively calm weather. A sea state of three or less, when there are few or no white caps to the waves, is preferable (see the section on *Weather, page 11*). In these conditions, large whales and leaping dolphins are often first noticed by the white water they create as they surface. The breath or 'blow' is another feature that can help to locate a large whale. Blows can often be seen several kilometres away and last for several seconds. Having observed an unusual splash or a possible blow, check with binoculars. It may be nothing, but more often than not, your first impressions will be correct and you will have found a cetacean!

Judging distance at sea

Judging distance at sea is not easy, especially when there are no landmarks for comparison, and this makes it difficult to gauge the size of a cetacean at the surface. The photograph above was taken looking forwards from a deck above the bridge of a cruiseferry.

The height of this deck is 35 m above sea level. For a person standing on this deck, the distance to the horizon is approximately 23 km. The superimposed lines show the distances from the front of the deck. Notice that the bands become narrower with increasing distance from the observer. Many land-based whale-watching locations are even higher above sea level and the effect is even more apparent. The advantage of a high viewing platform is that, due to the large field of view, it is often possible to see whale blows and dolphins splashing at several kilometres distance.

Watching from ferries

The European Atlantic is served by a plethora of ferry routes that offer passengers the opportunity to sail offshore in search of whales, dolphins and porpoises. Of these, several are now being used as platforms for conducting long-term research on cetacean distribution. Dedicated observers watching from a number of ferries that criss-cross the waters of the western English Channel, Celtic Sea and Irish Sea between England, France, Wales and Ireland regularly report whale and dolphin sightings. The most famous routes run from the south coast of England to the north coast of Spain; from Portsmouth and Bilbao and from Plymouth to Santander, crossing the Bay of Biscay. Both routes operate in all seasons and cover all four regions of the European Atlantic, providing access to some of the most productive waters for cetaceans in Europe.

The upper decks of ferries are generally the best place from which to look. Ferries generally travel at around 15–20 knots (nautical miles per hour), allowing cetaceans to be seen for a reasonable length of time. Sightings generally last from a few seconds to a few minutes, but if animals are seen ahead of the ship it may be possible to observe them for as long as 20 minutes. Both Biscay ferries have very high viewing platforms at over 30 metres above sea level. When first on board this seems

ridiculously high and it is easy to imagine that you will not see anything. Fortunately, cetaceans are quite large animals and this high vantage point often allows the observer to look down through the clear water to see whales and dolphins swimming below the surface. It is also a distinct advantage when wave height increases, as cetaceans are easily lost behind the waves at lower angles.

The height of the platform also makes it easy to see the formation and spacing of whole pods of dolphins and whales as they hunt, travel or rest at the surface, something that is often difficult to do from a small boat.

The other great advantage of whale-watching from such a large vessel is its stability. Large ships are so stable that in relatively calm conditions it is possible to use a tripod-mounted telescope and identify cetaceans at a considerable distance. Perhaps just as important to the discerning whale-watcher is the element of additional comfort provided – something which would certainly be lacking on a smaller vessel. Luxuries come in the form of Jacuzzis, swimming pools, cinemas, beauty salons, value-for-money shopping and good restaurants. However, if you do choose to dine during daylight hours, be sure to grab a window seat as many passengers have added the likes of Fin Whale, Cuvier's Beaked Whale and Striped Dolphin to the à la carte menu!

Watching from boats

Having a close encounter with a cetacean from a small boat is a very different experience to watching from a ferry or from land. Only when you are this close to the water can you meet the eye of a swimming dolphin as you gaze down from the bow, feel the spray of a whale's blow on your face, or listen to the wonderful sounds of dolphin whistles resonating through a boat's hull. For most of us, such opportunities are most readily available through one of an increasing number of commercial whale-watching boats operating in the region. Vessels are generally guided, providing excellent information on the local cetaceans and other marine life as you search. Try to choose an operator that has a clear ethical policy, including adhering to the local code of conduct for operating boats safely when approaching marine life.

Ferry-based whale-watchers in the Bay of Biscay.

Watching from land

Land-based watching can be an extremely rewarding alternative to taking a boat or ferry, especially as you can watch cetaceans without disturbing them, and it's absolutely free!

There are a number of excellent viewpoints around the European Atlantic coast, but it is possible to see cetaceans from just about anywhere, and always worth having a look. The best locations are headlands with a reasonably high vantage point, particularly those which extend to deep water. A good tip is to seek out lighthouses, which are often built on the most prominent headlands. Binoculars and often a telescope are useful tools for land-based watching. Most coastal areas are visited by cetaceans annually but some regions entertain a greater diversity and abundance of species than others. The best locations are blessed with some or all of the following: year-round presence of Bottlenose Dolphins and Harbour Porpoises; a good chance of Minke Whales during the summer, and occasional sightings of Short-beaked Common Dolphins, Risso's Dolphins, Long-finned Pilot Whales, Fin and Humpback Whales.

Land-based watching generally requires considerable patience and concentration while scanning the sea systematically. With experience you will discover the best weather, time of day and season to watch. Just taking in the magnificent view on a calm summer's day can make the trip to a coastal headland worthwhile. With so much sea to watch, you may be lucky enough to locate the blow of a large whale up to several kilometres away, pick out an unusual splash that turns out to be a breaching dolphin, or follow feeding seabirds flocks as they lead you to a group of Harbour Porpoises.

Whale-watchers gather at the cetacean hotspot of Galley Head in Ireland.

Principal whale-watching locations of the region

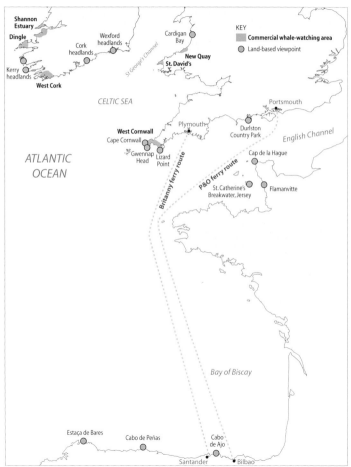

Cetacean Families

The cetaceans recorded in the European Atlantic can be divided into seven groups (which are broadly equivalent to families). This section aims to provide a brief overview of the key features which identify these groups, a broad knowledge of which is an essential first step towards identifying the different species at sea. It also provides certain historical background information.

RIGHT and BOWHEAD WHALES   Family: Balaenidae

North Atlantic Right Whale (page 68)

This family comprises some of the largest whales in the world. Four species are known: three right whales and the Bowhead Whale. Like the rorquals (Balaenopteridae), right and Bowhead Whales are very heavy, have a distinctive tall blow, and capture vast amounts of prey by filter-feeding. However, they differ from the rorquals in several respects. The body is much more stocky, with a broad back, completely

North Atlantic Right Whale.

lacking a dorsal fin. The head is huge, up to a quarter of the overall body length. The mouth consists of massive arched lips, within which lie very long baleen plates. Unlike the rorquals, the throat cannot be extended to take in more water but the baleen plates help to overcome this disadvantage by increasing the whale's filtering capacity. All four species are highly specialised plankton feeders. The North Atlantic Right Whale is the only member of the Balaenidae to occur in the European Atlantic. This was one of the first species to be hunted by early whalers as it frequently came close to shore, did not shy away from boats and floated when killed. It soon became known as the 'right' whale to catch, hence its present-day name. Hunting along the Biscay coast began around 1000 AD, and had spread across the ocean to Newfoundland by 1500 AD, continuing into the 20th century. Historically, the estimated population was 10,000–50,000 individuals but, although protected since 1949, this whale remains close to extinction. Less than 350 animals survive today, most occurring between Newfoundland and Florida.

RORQUAL WHALES   Family: Balaenopteridae

Blue Whale (page 66)
Fin Whale (page 28)
Sei Whale (page 26)
Humpback Whale (page 30)
Minke Whale (page 24)

With the exception of the Minke Whale, all rorqual whales are large to very large. They include the Blue Whale, the biggest animal ever to have lived on this planet. Sadly, rorquals formed the mainstay of the whaling industry during the 19th and 20th centuries, pushing several species to the brink of extinction. Following the international ban on whaling in 1986, some stocks have shown signs of a slow recovery, but many populations remain severely depleted. Rorquals have several features which distinguish them from the other groups.

Fin Whale.

SPERM WHALES Families: Physeteridae and Kogidae

Sperm Whale (page 32)
Pygmy Sperm Whale (page 70)
Dwarf Sperm Whale (page 71)

Although sperm whales include the largest and the smallest toothed whales from two different families, their basic body shapes are quite similar. Sperm whales are characterised by their bulky body, large, squarish head and narrow, underslung lower jaw lined with teeth. The single blowhole is positioned slightly to the left of the top of the head producing a blow which is angled forward and to the left when the whale exhales. All known species of sperm whale have been recorded in the European Atlantic. All sperm whales specialise in hunting deep-sea squid, and are therefore usually found far from land.

The name rorqual is derived from the Norwegian word '*rorvhal*' which refers to the series of throat grooves that extend underneath the lower jaw. Rorquals do not possess teeth, instead having a series of comb-like structures called baleen plates which hang from the upper jaw. They feed by opening their cavernous jaws as they swim along, and expanding their throat grooves, vastly increasing the volume of water held within. When the mouth is finally closed, water is strained through the baleen plates leaving quantities of small fish and zooplankton trapped inside. Despite their enormous size, rorquals are streamlined and capable of considerable speed. They have large, flattened heads with a centrally-located twin blowhole. The larger species can exhale a vertical blow several metres high. Most members of this group migrate long distances between their cold water summer feeding grounds and warm water winter breeding grounds.
Five species of rorqual are known to occur in the European Atlantic.

Sperm Whale (travelling left to right).

BEAKED WHALES

Beaked whales are unique amongst the medium-sized whales in possessing a distinctive protruding jaw, or beak. The forehead is often bulbous, the body relatively stream-lined (often showing scarring) and the dorsal fin is smallish, located two-thirds of the way towards the tail. The centre of the trailing edge of the tail has no notch. Due to their similar body shapes and coloration, most beaked whales, particularly those in the genus *Mesoplodon* (pages 40–41), are notoriously difficult to identify. The main difference between the species in this genus is the position of protruding teeth which erupt only in mature males. The observer will almost certainly have to note, and ideally photograph, this feature to prove identification.

Six of the 20 known species of beaked whale have occurred in the European Atlantic. However, because of their deep-water habits and unobtrusive behaviour almost nothing is known about any of them. Beaked whales are capable of diving for considerable periods of time, and to great depths, in search of their squid and deep-sea fish prey.

Family: Ziphiidae

Northern Bottlenose Whale (page 36)
Cuvier's Beaked Whale (page 38)
Sowerby's Beaked Whale (page 42)
Gervais' Beaked Whale (page 72)
True's Beaked Whale (page 43)
Blainville's Beaked Whale (page 73)

Gervais' Beaked Whale.

BLACKFISH

Long-finned Pilot Whale (page 46)
Short-finned Pilot Whale (page 74)
Killer Whale or Orca (page 44)
False Killer Whale (page 75)
Pygmy Killer Whale (page 76)
Melon-headed Whale (page 77)

Family: Delphinidae

Blackfish include the largest members of the dolphin family, and, as the name suggests, they are predominantly black in colour. The dorsal fin is relatively large and prominent. A low, bushy blow can usually be seen emanating from the single blowhole on surfacing. The jaws are rounded, containing many well-developed conical teeth, and the beak is small or lacking. Like other dolphins, blackfish are highly social animals, often living their entire lives in discrete family groups. They are efficient, pack-hunting predators, capable of considerable speed and exhibit a range of interesting behaviour such as spy-hopping, tail-slapping and breaching. They prey on a great variety of fish, squid and other sea life, including marine mammals. All six species of blackfish have occurred in the European Atlantic.

Long-finned Pilot Whale.

Two new species of beaked whale have been discovered in the last twenty years, indicative of the enormous amount still to be learnt about the distribution, ecology and identification of this mysterious family.

DOLPHINS
Family: Delphinidae

White-beaked Dolphin (page 60)
Atlantic White-sided Dolphin (page 58)
Bottlenose Dolphin (page 54)
Risso's Dolphin (page 56)
Short-beaked Common Dolphin (page 50)
Striped Dolphin (page 52)
Rough-toothed Dolphin (page 78)
Fraser's Dolphin (page 79)

Dolphins are smaller than most whale species, have slim, streamlined bodies and, in most cases, tall, prominent and centrally-placed dorsal fins. Their bodies show a variety of patterns and colours which are often key to their identification. The heads of most species taper gently to a prominent beak which contains many sharp teeth. They are generally social animals, often occurring in large groups and are capable of great speed and spectacular acrobatics. They prey upon a wide variety of squid, fish and other marine life. Eight of the 31 known species of dolphin have occurred in the European Atlantic.

Short-beaked Common Dolphin.

PORPOISES
Family: Phocoenidae

Harbour Porpoise (page 62)

Porpoises are the smallest of all cetaceans, reaching 1·2–2·0 metres. They do not have a prominent beak and the body, though streamlined, is quite robust. Most species are timid, rarely performing the kinds of acrobatics commonly associated with dolphins. They frequent inshore waters, feeding mostly on small fish. Unfortunately, the pressures on many coastal populations through overfishing, accidental capture in fishing nets, and pollution have caused significant declines in several species to the extent that they are currently considered threatened. Of the six known species of porpoise, only one occurs in the European Atlantic.

Harbour Porpoise.

Harbour Porpoise 1·5 m

Common Dolphin 2 m

Humpback Whale
11–15 m
page 30

Killer Whale (female) 5–6 m

Sperm Whale
11–18 m
page 32

SCALE

All the cetacean species illustrated on pages 22 and 23 are depicted at a scale of 1:125 or 0·8% of life size. The smaller cetaceans (above) are included to illustrate the immense difference in size between them and the great whales, something that is often difficult to appreciate when viewing a great whale as it surfaces to reveal only a small proportion of its back at any one time.

North Atlantic Right Whale
11–18 m
page 68

Minke Whale

Balaenoptera acutorostrata

F: Petit rorqual
ESP: Rorcual aliblanco

Adult length:	7–10 m
Group size:	1–2, occasionally more
Breaching:	Variable angles
Deep dive:	Tail flukes not raised
Blow:	Small, vertical and bushy, but usually not visible

BEHAVIOUR: Minke Whales are usually solitary animals, sometimes seen in pairs or small groups, but rarely in larger aggregations. They can be very elusive and hard to follow when feeding actively but are often quite inquisitive, occasionally approaching passing boats and turning on their sides as they swim past or lifting their heads clear of the surface. Breaching is regularly recorded, particularly during periods of rough weather.

STATUS AND DISTRIBUTION: Minke Whales have a worldwide distribution, but they appear to be more common in cooler waters than in the tropics. They are frequently seen during the summer and autumn in coastal waters around the British Isles, the Faroes, Iceland and western Norway. In the European Atlantic, the Minke Whale is the most regularly recorded whale in shelf waters, occurring in greatest numbers from northern France off the Brittany coast, north and west to the western English Channel, Southern Irish Sea and west to Ireland, particularly during the summer months. Most sightings are of single animals, although small aggregations and mother-calf pairs have also been seen.

J F M A M J J A S O N D

SURFACING SEQUENCE

DIRECTION OF TRAVEL →

The surfacing roll of a Minke Whale is relatively fast. The inconspicuous blow appears at the same time as the head and snout which emerges at an angle ...

... the top of the head and back forward of the dorsal fin is the next to show, sometimes with the blow ...

... and then the animal rolls quickly appears ...

... then the dorsal fin appears ...

... exposing the tail-stock which then sinks beneath the surface (compare with Sei Whale page 26).

Observing a Minke Whale at sea can be quite hard as it may only roll once or twice before disappearing!

SIMILAR SPECIES: Confusion is most likely with the larger rorquals, particularly Sei (*page 26*) and Fin Whales (*page 28*). However, the much smaller size and lack of a tall blow should distinguish it from these species. Minke Whale is also easily confused with the larger beaked whales (*pages 36* and *38*), which are similar in size and shape although the Minke Whale's pointed rostrum with a central ridge is diagnostic. Beaked whales tend to occur in much deeper waters than Minke Whales, although location alone should not be used to identify this species.

SIMILAR SPECIES: BACK AND FIN PROFILES

Northern Bottlenose

Minke

Sei

Fin

Minke Whale has a slender, streamlined body with a pointed rostrum which is bisected by a single longitudinal ridge beginning in front of the blowholes.

The dorsal fin is tall and falcate and placed about two-thirds along the back.

IDENTIFICATION:
The Minke Whale is the smallest baleen whale to be found in the European Atlantic, and the only whale without a highly visible blow that occurs regularly in shallow water and close to land.

The upperparts are dark grey, lightening to white on the belly and the underside of the flippers.

A distinctive feature, although only conspicuous at close range, is a diagonal white band on the upper surface of the flipper. When diving, they do not raise their tail flukes clear of the surface.

Minke Whales have an inconspicuous vertical blow, which is produced at almost the same time as the dorsal fin appears.

A Minke Whale rolls to dive.

Sei Whale

Balaenoptera borealis

F: Rorqual boreal
ESP: Rorcual boreal

Adult length:	12–16 m
Group size:	1–2, sometimes more
Breaching:	Seldom, generally rising at a low angle
Deep dive:	Does not raise tail flukes
Blow:	Tall, thin and vertical, less robust than Fin Whale

BEHAVIOUR: Sei Whales tend to travel alone or in pairs. Unlike other rorquals, they often skim-feed just below the surface. When feeding in this manner they can be quite unobtrusive, with only the occasional blow and top of the dorsal fin being visible. They are very fast swimmers, but do not appear to be as acrobatic as either Fin or Minke Whales at the surface.

When a Sei Whale surfaces, the blow and fin *usually* appear at (or almost at) the same time (compare with Fin Whale page 28) …

… then the animal rolls slowly at a shallow angle with the dorsal fin prominent …

… before rolling and sinking …

… with the dorsal fin being the last part visible.

STATUS AND DISTRIBUTION:

Sei Whale has a cosmopolitan distribution, but appears to favour temperate to sub-polar waters. They are generally restricted to deep, pelagic waters and are rarely found near coasts. In Europe, small numbers are seen to the west of Britain and Ireland, and north to Iceland and northern Norway. This species was a regular visitor to the Bay of Biscay in the late 1990s, but since then it has become much less frequent, with just a handful of sightings each year. The majority of records are during July, August and September and occur beyond the shelf-edge over deep water. Sightings in other seasons are rare, but one notable exception involved two individuals engaged in skim-feeding at the surface in the Southern Bay in March 2001, with several Short-beaked Common Dolphins swimming alongside. There are one or two sightings of this species in coastal waters of the European Atlantic annually.

J F M A M J J A S O N D

SURFACING SEQUENCE

DIRECTION OF TRAVEL →

SIMILAR SPECIES: This species can easily be confused with both Fin (*page 28*) and Minke (*page 24*) Whales. In general, a combination of features should be used to separate these species. Compared with Sei Whale, Fin Whale usually has a higher blow which appears well before the dorsal fin, which itself is lower and more sloping. Unlike Fin Whales, Sei Whales do not usually arch their tail-stock prior to a deep dive. At close range, the asymmetrical head pattern of Fin Whale is diagnostic. Minke Whales are distinguished from Sei Whales by their smaller size and lack of a prominent blow.

IDENTIFICATION: Sei Whale is a large, dark whale with an upright dorsal fin positioned two-thirds of the way along the back.

Similar in form and colour to a small Fin Whale, it has an erect, sickle-shaped dorsal fin which appears proportionally larger than that of Fin Whale.

The blow is quite tall, but not as prominent as in Blue or Fin Whale.

A median ridge on the head extends from the blowholes almost to the tip of the rostrum.

While feeding or travelling, the back and dorsal fin are usually visible for a longer period compared with other rorquals.

SIMILAR SPECIES: FIN PROFILES
Sei   Fin   Minke

When surfacing, the blowholes and fin appear almost simultaneously.

Sei Whale often appears to sink, rather than roll, and the dorsal fin is the last part of the animal that is above the surface.

Fin Whale

Balaenoptera physalus

F: Rorqual commun
ESP: Rorcual común

Adult length:	18–26 m
Group size:	1–5, sometimes more
Breaching:	Variable angles; huge splash
Deep dive:	Tail flukes not raised
Blow:	Tall column to 8 m, thicker and higher than Sei but smaller than Blue Whale

BEHAVIOUR: Fin Whales are usually observed singly or in pairs, sometimes in small pods, and occasionally in large aggregations of up to 100 animals where food is plentiful. They are very fast swimmers and have regularly been recorded breaching and lunge-feeding.

STATUS AND DISTRIBUTION: Fin Whales are widely distributed throughout the world, but are more common in sub-polar and cool temperate waters in the summer months, moving to warmer latitudes in the winter to mate and calve. The Fin Whale is recorded regularly in the region, particularly in deep waters over and beyond the edge of the continental shelf. The Bay of Biscay appears to be principally a late-summer feeding ground for this species, with most sightings between July and October with a peak in August. Sightings usually involve lone animals or pairs, although larger aggregations of up to 30 individuals sometimes occur on productive feeding grounds. In some years the whales feed at depth, while in others they prey upon large volumes of small silvery fish that are visible at the surface. To feed on these fish, the whales often swim on their side with one tail fluke raised clear of the water, sometimes engaging in spectacular surface lunges with their jaws wide open. Throughout the autumn, winter and spring, a few move into shallow shelf waters, such as off Cornwall and south-west Ireland, to feed on shoaling fish.

SURFACING SEQUENCE

→ DIRECTION OF TRAVEL

The blow appears first ...

... then the first portion of the back

... and then the dorsal fin appears.

A travelling whale then rolls over and repeats the sequence ...

... unless it goes into a deep dive, when the fin and tail-stock create a characteristic outline.

IDENTIFICATION:
A large species, second only in size to Blue Whale. Fin Whale is a relatively slender, streamlined animal with a pointed head and a prominent dorsal fin positioned three-quarters of the way along the back.

SIMILAR SPECIES: At long range confusion is likely with both Sei (*page 26*) and Blue Whales (*page 66*). Sei Whales generally have a comparatively weaker blow, which, on surfacing, appears simultaneously with a more prominent sickle-shaped dorsal fin. Also, when diving, Fin Whales are more likely to arch their tail-stock. Blue Whales differ in having a small dorsal fin, paler coloration and a mottled body pattern. Unlike Blue Whales, Fin Whales only very rarely raise their tail flukes on diving.

The upperside of the body is dark grey, often with pale chevrons behind the blowholes.

At close range, the asymmetrical pigmentation of the lower jaw is diagnostic – the left lower jaw being dark (BELOW), the right being white (LEFT).

The tall, columnar blow takes several seconds to dissipate.

SIMILAR SPECIES: FIN PROFILES

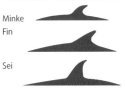

Minke
Fin
Sei

A Fin Whale starts a deep dive.

Typical surfacing Fin Whale.

Humpback Whale

Megaptera novaeangliae

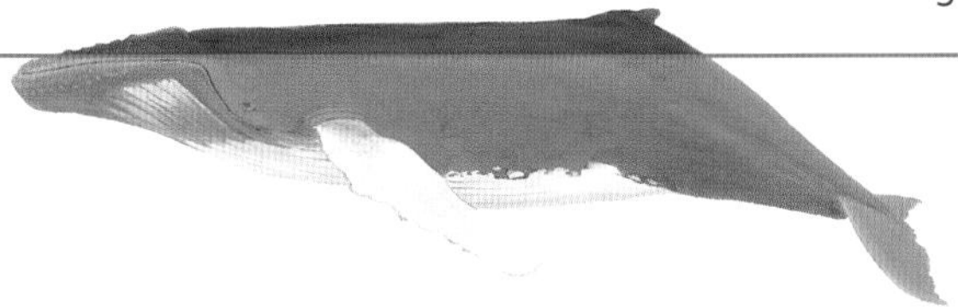

F: Mégaptère
ESP: Jorobada

Adult length:	11–15 m
Group size:	1–3, sometimes more
Breaching:	Frequently, usually landing on back
Deep dive:	Body generally arches high before broad tail flukes are raised to reveal variable pale underside to tail
Blow:	Variable; tall, vertical and bushy

BEHAVIOUR: Humpbacks can be very demonstrative and are regularly observed breaching (see *page 14*), lob-tailing, and slapping their long flippers on the surface of the water. They are very vocal and are known as the singing whales. Although they usually travel singly or in very small pods, large congregations of animals commonly occur on the summer feeding grounds.

STATUS AND DISTRIBUTION:

Humpback Whales are found throughout the world. During the summer months they feed in polar and cool temperate waters, migrating in the winter to sub-tropical and tropical latitudes, where they mate and calve. This species was heavily exploited by the whaling industry, but it now appears to be making a good recovery in many parts of the world. In European waters, Humpbacks appear to be making a slow recovery, with a steady increase in the number of animals seen in shelf waters off south-west Ireland, and to a lesser extent, the UK, in recent years. Despite this, the Humpback Whale remains an extremely rare animal in the Bay of Biscay, with only a handful of reports from ferries over the last decade and no confirmed sightings. This may be due to the migration route of most Humpback Whales summering in northern Europe, which move south-west to the Caribbean for the winter, and are therefore unlikely to occur as far south and east as the Bay of Biscay with any regularity.

J F M A M J J A S O N D

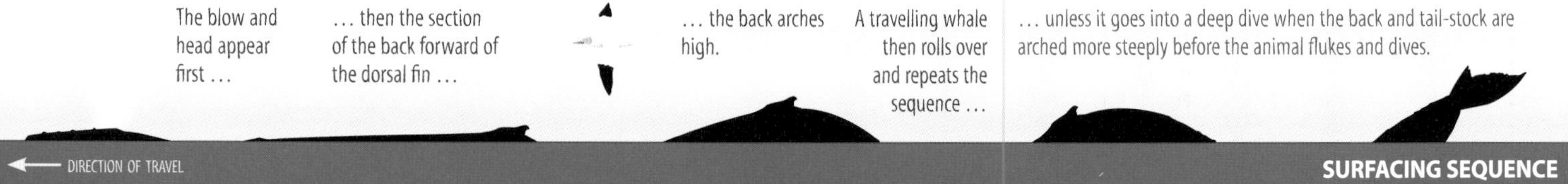

SURFACING SEQUENCE

The blow and head appear first ... then the section of the back forward of the dorsal fin ... the back arches high ... A travelling whale then rolls over and repeats the sequence ... unless it goes into a deep dive when the back and tail-stock are arched more steeply before the animal flukes and dives.

DIRECTION OF TRAVEL →

IDENTIFICATION:

One of the most distinctive features is the extremely long, mostly white, pectoral flippers.

SIMILAR SPECIES: Although similar to several other rorquals, Sperm Whale (*page 32*) and North Atlantic Right Whale (*page 68*) in size, the distinctive flippers, fluke pattern and surfacing sequence separates the Humpback Whale from all other species.

The trailing edge of the flukes is always irregular and the patterning on the undersides can be used to identify individuals.

The blow is dense and mushroom-shaped.

Before a sounding dive, the body usually arches steeply before the tail flukes are raised high above the surface.

The Humpback Whale is a large, stocky whale, black or dark grey in colour with a broad head covered in fleshy tubercles.

The dorsal fin is usually short and stubby.

Sperm Whale

Physeter macrocephalus

F: Cachalot macrocephal
ESP: Cachalote

Adult length:	11–18 m
Group size:	1–20, sometimes more
Breaching:	Leaps vertically
Deep dive:	Often arches body; usually raises dark, broad, triangular tail flukes vertically
Blow:	Angled forwards and to the left

BEHAVIOUR: Sperm Whales are gregarious, often travelling in pods of up to 20 or more and can be quite demonstrative, breaching regularly. They are usually encountered either as mixed pods of females and immature males, or lone males/bachelor groups. This species is renowned for diving to tremendous depths to capture deep-water fish, sharks and squid, including Giant Squid.

STATUS AND DISTRIBUTION:

The Sperm Whale is widely distributed in deep, offshore waters throughout the world. The mixed female/immature male pods usually remain in tropical or sub-tropical waters throughout the year. Males travel to higher latitudes, sometimes as far as the Arctic ice-edge, to feed, and return to join the groups of females for the breeding season. Sperm Whales are encountered in low numbers in the European Atlantic regularly. Some years groups of females with calves have been noted, possibly when water temperatures are higher than normal which suggests that the Bay of Biscay forms the northern limit of their range. In other years, sightings are generally of smaller groups or solitary, larger animals, presumed to be mature males. Some of these animals may be present throughout the summer, although there is a marked increase in sightings between August and October, probably males on their southward migration to the breeding grounds.

J F M A M J J A S O N D

SURFACING SEQUENCE

The distinctive angled blow appears first …

… then slowly the head, back and dorsal hump emerge and the whale may remain in this profile, either resting motionless or travelling very slowly …

… until it dives, rolling forward in a steep arch …

… before the tail-stock rises …

… and the animal lifts its flukes vertically as it starts its dive.

IDENTIFICATION: The largest of the toothed whales: mature males average about 15 m in length; females about 10 m. The body is robust, with a massive, blunt head that comprises almost one-third of the animal's total length. The lower jaw is long, narrow and inconspicuous beneath the head.

SIMILAR SPECIES: Sperm Whale overlaps with the largest whales in size and, like Humpback (*page 30*), Blue (*page 66*) and North Atlantic Right Whale (*page 68*), raises its flukes before a deep dive. However, all baleen whales are distinguished by their vertical blow, and all rorquals have a distinctive dorsal fin.

Sperm Whales are normally slate grey or brown and the skin is wrinkled.

There is no dorsal fin, only a small hump, behind which can be a series of 'knuckles' running down to the tail.

The flukes, which are generally raised vertically prior to a deep dive, are large and triangular, with smooth edges and a deep central notch.

The distinctive blow is low and 'forward-angled' due to the offset position of the blowhole.

A single blowhole is located on the left side of the front of the head.

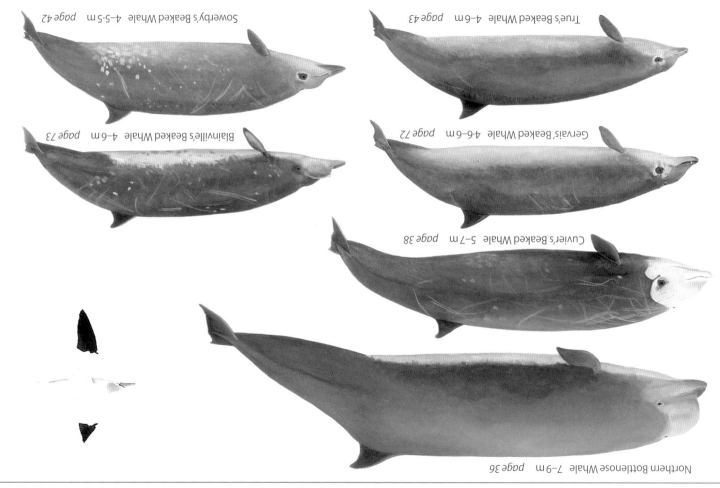

The Blackfish

Long-finned Pilot Whale
3·5–6·5 m *page 46*

Killer Whale or Orca (female)
5–9 m *page 44*

Short-finned Pilot Whale
3·5–6·5 m *page 74*

Pygmy Killer Whale
2·0–2·7 m *page 76*

False Killer Whale
4–6 m *page 75*

Melon-headed Whale
2·0–2·7 m *page 77*

Northern Bottlenose Whale

Hyperoodon ampullatus

F: Hyperoodon boreal
ESP: Zifio boreal

Adult length:	7–9 m
Group size:	1–10
Breaching:	Lifts entire body out of the water, sometimes several times in a sequence
Deep dive:	Tail flukes not raised
Blow:	Low and bushy, slightly angled forwards

BEHAVIOUR: Northern Bottlenose Whales are generally seen in pods moving slowly at the surface. Breaching has been observed occasionally.

STATUS AND DISTRIBUTION: This is the only species of beaked whale to have been hunted commercially in the North Atlantic.

with the depleted population finally being afforded protection in 1977. Northern Bottlenose Whales are restricted to the deep offshore waters of the northern North Atlantic, with the Bay of Biscay being towards the southernmost extent of their range. The Biscay population of this rare and little-studied cetacean seems to be centred over deep-water canyons of the southern Bay, although it has been sighted further north. Northern Bottlenose Whales were sighted regularly in the Bay of Biscay in the late 1990s, but sightings since then have been restricted to a handful every year. In recent years, this species has also been seen increasingly around the coasts of the UK and Ireland, although it remains very rare.

J F M A M J J A S O N D

SURFACING SEQUENCE

→ DIRECTION OF TRAVEL

The distinctive bulbous melon appears first (sometimes at a steep enough angle to reveal the beak) with the blow visible in calm conditions …

… then the back and dorsal fin with the top of the head usually still visible …

… and then the animal rolls languidly …

… and submerges with the tail-stock slightly arched.

IDENTIFICATION: Although Northern Bottlenose Whale is the largest species of beaked whale in the European Atlantic. it can only be reliably identified by seeing the large bulbous head as the animal breaks surface.

SIMILAR SPECIES: Head-shape separates Northern Bottlenose Whale from both the similar Cuvier's Beaked Whale (*page 38*) and Minke Whale (*page 24*). Pod sizes of greater than two largely rule out Minke Whale.

At close range, the rounded forehead can be seen to overhang slightly the short, protruding beak. Mature individuals often have a paler head.

Northern Bottlenose Whale is medium-sized, uniform grey to brown in colour, and has a smallish, falcate dorsal fin situated two-thirds along the back.

Cuvier's Beaked Whale

Ziphius cavirostris

F: Ziphius
ESP: Zifio común

Adult length:	5–7 m
Group size:	1–12
Breaching:	Leaps almost vertically
Deep dive:	Tail flukes not raised
Blow:	Low and bushy, slightly angled forwards

BEHAVIOUR: This scarce and poorly-known whale has been sighted regularly in the region and recent observations have led to some new discoveries and exciting glimpses into its life history. Like elephants, it appears that male Cuvier's Beaked Whales use their tusks to fight each other to gain access to groups of mature females in order to mate. Also of interest is that sightings of groups of females with calves have not included males, possibly because mothers or calves may be injured if they remain in the vicinity of fighting males. Apparently inquisitive animals have been recorded many times from the ferries operating in the Bay of Biscay, often surfacing very close by, logging or moving slowly within a few tens of metres of the ship.

STATUS AND DISTRIBUTION: Stranding records suggest that Cuvier's Beaked Whale is cosmopolitan in temperate and tropical waters. In Europe, this species is occasionally sighted in deep waters as far north as Ireland. However, it is regularly seen throughout the deep waters of the Bay of Biscay, with the majority of records coming from the canyons of the Southern Bay. Indeed, Cuvier's Beaked Whale has been sighted on up to 75% of Biscay ferry crossings during some summers, making these waters the most reliable location to see this species anywhere in the world.

SURFACING SEQUENCE

→ DIRECTION OF TRAVEL

The melon and head appear first (sometimes at a steep enough angle to reveal the beak), with the blow sometimes visible in calm conditions ...

... then the back and dorsal fin with the top of the head usually still visible ...

... and then the animal rolls languidly ...

... and submerges with the tail-stock slightly arched.

IDENTIFICATION: Cuvier's Beaked Whale is one of the largest and most robust of the beaked whales, best identified by its size, coloration and head-shape. The shape of the head and beak has been described as being like that of a goose, the forehead sloping gently to the tip of the short, but distinct, beak. Mature females and immature animals of both sexes lack protruding teeth. However, mature males develop a pair of teeth that protrude upwards from the tip of the lower jaw. The blow is sometimes visible.

SIMILAR SPECIES: Cuvier's Beaked Whale is similar in size and shape to Minke Whale (*page 24*) and other beaked whales, but the head-shape distinguishes it from these species. Without the head-shape and pale coloration being seen, Cuvier's Beaked Whale cannot be reliably identified.

A heavily scarred mature male

Most observations do not provide views of the beak, but the gently-curving forehead is quite unlike the bulbous melon which is characteristic of some beaked whales.

Although young calves have all-dark bodies, immature animals generally develop pale heads, and, with age, this may extend along the back as far as the dorsal fin. The upper body is often heavily scarred in mature males, whereas females generally show little or no scarring.

A medium-sized grey-brown to brown whale with a longish back and a small falcate dorsal fin situated two-thirds along the back.

Identifying *Mesopolodon* beaked whales: the toughest challenge of all

Six species of beaked whale have occurred in the European Atlantic. All of them are challenging to identify at sea, although the Northern Bottlenose Whale (*page 36*) and Cuvier's Beaked Whale (*page 38*) are distinctive given good views.

By far the most difficult to identify of all the cetaceans in the European Atlantic are the *Mesopolodon* beaked whales, of which four are known to occur.

As in other parts of the world, the vast majority of records are of strandings, where a positive identification is usually possible.

The sighting of a *Mesopolodon* beaked whale at sea is generally considered to be an exceptional event due to the shy habits, unobtrusive behaviour and apparent rarity of all species throughout their known range in deep, offshore waters. It is, therefore, especially exciting that these whales are being sighted on several occasions annually in the southern Bay of Biscay.

Unfortunately, because *Mesopolodon* beaked whales are so difficult to identify, most sightings are not recorded to species. There are still only a handful of photographs taken at sea, and photographs of stranded animals reveal little about their true coloration and nothing of their surfacing behaviour.

The plate below illustrates the diagnostic features which are required to confirm identification of each species, based on current knowledge. It is important to note that, although the plate shows the whales raising their heads clear of the water, it is not known whether all of the species regularly do this when surfacing (although some have been seen to do so).

A close view of the head of an adult male is crucial to the positive identification of Gervais', True's and Sowerby's Beaked Whales. This is because only the mature males show a pair of protruding teeth, the only feature which reliably separates the three species. Fortunately, these teeth are situated at different positions on each species, as detailed below. Blainville's Beaked Whales are perhaps slightly easier to identify, although the body shape and coloration can be almost identical to the other species; adults generally show an arched lower jaw, from which a pair of protruding teeth erupt in males. In adult females, the tip of the jaw is often white.

Although the surfacing sequence below depicts a Sowerby's Beaked Whale (*page 42*), the surfacing sequences of all the *Mesopolodon* beaked whales are presumed to be similar.

SURFACING SEQUENCE

→ DIRECTION OF TRAVEL

The melon and head appear first (sometimes at a steep enough angle to reveal the beak) with the blow visible in calm conditions ...

... then the back and dorsal fin with the top of the head usually still visible ...

... and then the animal rolls smoothly ...

... and submerges with the tail-stock slightly arched.

40

Mesoplodon Beaked Whales

IMPORTANT NOTE

The lack of confirmed sightings of *Mesoplodon* beaked whales means that there is little reference material on which to base illustrations.

The illustrations shown here and throughout the book are based on limited photographic and video footage from the following sources:

True's Beaked Whale:
Surface – Bay of Biscay (two encounters).

Gervais' Beaked Whale:
Surface – Canary Islands;
Strandings – Florida, Cayman Islands.

Sowerby's Beaked Whale:
Surface – Bay of Biscay, Canary Islands;
Underwater – Maldives;
Strandings – Newfoundland, Scotland.

Blainville's Beaked Whale:
Surface – Canary Islands;
Underwater – Maldives;
Stranding – North Carolina.

However, there may be a large variance in appearance, depending on factors such as location, age and gender.

Notes relating to these variations are detailed alongside the illustrations.

TRUE'S BEAKED WHALE (male) (*page 43*)
FOREHEAD: Round taper to short, stubby beak.
TEETH: At tip of lower jaw.
COLORATION: One source image shows a dark-tipped beak and virtually all-dark body (see *page 34*), a second source image shows an individual with dark/light pattern similar to Gervais' Beaked Whale.

SOWERBY'S BEAKED WHALE (male) (*page 42*)
FOREHEAD: Roundish taper to a long, slender beak.
TEETH: Midway along lower jaw.
COLORATION: Sources show animals ranging from individuals with typical dark/light pattern to some with all dark bodies (see *page 34*) and beaks ranging from dark-tipped to all dark.

BLAINVILLE'S BEAKED WHALE (male) (*page 73*)
FOREHEAD: Raised, arched jawline.
TEETH: Erupt from arching lower jaw,
COLORATION: Sources show some animals with dark/light pattern, others with all dark bodies (see *page 34*) and beaks ranging from dark-tipped to all-dark.

GERVAIS' BEAKED WHALE (male) (*page 72*)
FOREHEAD: Smooth taper to a shortish, slender beak.
TEETH: one third distance from tip of lower jaw.
COLORATION: References (3 images) show a consistent light/dark pattern (see *page 34*).

SOWERBY'S BEAKED WHALE (female)
FOREHEAD: Flatter taper than male to a long, slender beak.
TEETH: None.
COLORATION: References show animals with classic *Mesoplodon* dark/light pattern and dark eye.

BLAINVILLE'S BEAKED WHALE (female)
FOREHEAD: Raised jawline.
TEETH: None.
COLORATION: References show animals with either brown or grey bodies, and white or all brown lower jaws.

Sowerby's Beaked Whale
Mesoplodon bidens

F: Mésoplodon de Sowerby
ESP: Zifio de Sowerby

Adult length:	4–5.5 m
Group size:	1–8
Breaching:	Leaps almost vertically
Blow:	Small, bushy, slightly angled forwards

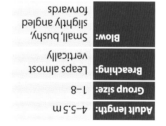

BEHAVIOUR: Sowerby's Beaked Whales are believed to be unobtrusive in their behaviour and generally shy of boats. However, breaching has been recorded on several occasions.

STATUS AND DISTRIBUTION: Sowerby's Beaked Whale is restricted to the northern North Atlantic. Most European sightings are at latitudes between the Canary Islands and the Arctic Circle, with the deep waters to the west of the UK and Ireland considered to be the centre of the species' regular range. In the Bay of Biscay, sightings from ferries and strandings on beaches have occurred in roughly equal measure, totalling just a handful of records annually. The status of Sowerby's Beaked Whale in the European Atlantic remains poorly known, but these few records suggest that it is the most frequently occurring *Mesoplodon* beaked whale in the region.

SIMILAR SPECIES: Sowerby's Beaked Whale is almost identical in size, shape and coloration to Gervais' (page 72), True's (below) and Blainville's Beaked Whales (page 73).

J F M A M J J A S O N D

IDENTIFICATION: Only rarely sighted at sea, Sowerby's Beaked Whale is known from a small number of encounters. It is a small, slim whale, reminiscent in shape to an elongated dolphin. The upper surface is uniformly brown to grey, becoming paler towards the belly, and the eyes are often encircled by a dark patch. As with all beaked whales, identification at sea is dependent upon observation of the head. The rounded forehead tapers to a slender beak, the length of which, although variable, is longer than the other beaked whales. The beak often breaks the water at a steep angle on surfacing and the observer must concentrate on this feature in an attempt to clinch identification. The only known diagnostic feature for this species at sea is the two flattened teeth that protrude mid-way along the lower jaw in adult males (see page 41).

True's Beaked Whale
Mesoplodon mirus

F: Mésoplodon de True
ESP: Zifio de True

BEHAVIOUR: With only a handful of sightings ever made at sea, the behaviour of True's Beaked Whale remains relatively unknown. Although the species has been observed breaching vertically over twenty times in succession, it seems likely that it generally moves unobtrusively at the surface and avoids boats.

STATUS AND DISTRIBUTION: The vast majority of stranding records for True's Beaked Whale are from the North Atlantic, although there are some from South Africa and Australia. In Europe, a few strandings have occurred on the coasts of Spain, France, Ireland and the Canary Islands. There have also been a handful of probable sightings of this species at sea in the Bay of Biscay over the last decade, but only one has been confirmed beyond doubt: on 9th July 2001, a mature male was photographed breaching 24 times in the southern Bay. This sighting, which was observed by over 100 ferry passengers, constitutes the first confirmed record of this species at sea anywhere in the world.

Adult length:	4–6 m
Group size:	1–6
Breaching:	Leaps almost vertically
Blow:	Unknown

J F M A M J J A S O N D

IDENTIFICATION: True's Beaked Whale is known from only a handful of stranding records and encounters at sea. There is consequently much to learn about its identification, behaviour and ecology. As with other *Mesoplodon* beaked whales, the key to identification is a good enough view of the head to ascertain the position of protruding teeth in an adult male. The head in front of the blowhole is distinctly bulbous, with a dark patch around the eye and a short, stubby beak. In adult males, a pair of teeth protrude from the tip of the lower jaw (see *page 41*). Like other *Mesoplodon* beaked whales, this is a slim whale with a spindle-shaped body. The back is grey, though some individuals may show a darker spinal line and dorsal fin. The dorsal fin, which may appear falcate to triangular, is situated two-thirds of the way along the back.

SIMILAR SPECIES: True's Beaked Whale is extremely similar to other *Mesoplodon* beaked whales, making identification at sea almost impossible without good views of the head.

Killer Whale or Orca

Orcinus orca
F: Orque
ESP: Orca

Adult length:	5–9m
Group size:	Family groups of 2–30
Breaching:	Leaps vertically
Deep dive:	Tail flukes not raised
Blow:	Tall and bushy

BEHAVIOUR: In some regions of Europe, Killer Whales hunt fish such as Herring, Cod, Mackerel and Salmon, but elsewhere they hunt seals, and may attack other whales and dolphins. Their appetite for hunting large baleen whales in the southern oceans is why whalers first named them whale killers, hence the alternative present day name Killer Whale. There is some evidence that Killer Whales in the European Atlantic may hunt other marine mammals as part of their regular diet, including two encounters in the Bay of Biscay in August; the first when a pod of five Killer Whales was seen in association with four Fin Whales, and possibly targeting a Fin Whale calf that was with the group; the second when two bull Killer Whales were seen in apparent pursuit of two adult pilot whales. Elsewhere in the region there are also records of live strandings of pilot whales and Short-beaked Common Dolphins, with Killer Whales in the vicinity.

STATUS AND DISTRIBUTION:

Records in the European Atlantic suggest that Killer Whales occur in very small numbers in all water depths. Although rare along the Biscay coasts of France and Spain and in the English Channel, Killer Whales are regularly sighted close to shore off south-west Ireland during the summer months, and are occasionally seen off Cornwall, England, and in the Irish Sea. In the Bay of Biscay, they are only sighted on a handful of occasions each year.

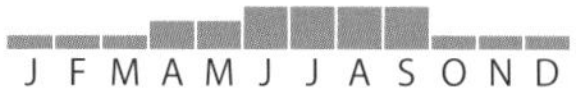

J F M A M J J A S O N D

SURFACING SEQUENCE

DIRECTION OF TRAVEL →

The blow appears at the same time as the head and dorsal fin ...

... then the first portion of the back ...

... and then the dorsal fin shows prominently ...

... before the tail-stock is arched prior to diving.

Female/immature

Adult male.

Dorsal fin shape varies with age and gender, with adult males unmistakable.

SIMILAR SPECIES: In poor light, or at long range, when the white eye patch is not visible, it is possible to mistake female or immature Killer Whales for other members of the blackfish family or large dolphins such as Risso's Dolphin (*page 56*).

Killer Whales spend most of their lives in discrete family groups of two or more animals, so most sightings involve individuals with different shaped dorsal fins. These powerful predators are capable of swimming at high speed and regularly breach.

IDENTIFICATION: Despite its name, the Killer Whale is actually the world's largest dolphin, and perhaps the most striking and familiar cetacean of all. The stocky, sleek, black body, white eye patch, and prominent dorsal fin are unmistakable. Most animals show a clear grey 'saddle' patch behind the dorsal fin.

Fin shape and size varies depending on the age and sex of the individual.

Adult males are the most distinctive, with broad, triangular dorsal fins reaching up to 2 m in height.

Females and immature animals are significantly shorter in body length, with smaller, sickle-shaped dorsal fins.

Long-finned Pilot Whale

Globicephala melas

F: Globicéphale noir
ESP: Calderón común

Adult length:	3.5–6.5 m
Group size:	Family groups of 2–50
Breaching:	Various angles
Blow:	Distinctly bushy

BEHAVIOUR: These whales are usually seen in family groups. They feed mostly at night, spending much of the day travelling leisurely or logging at the surface. They are inquisitive animals and often appear close to ships, sometimes spy-hopping and tail-slapping (see *page 14*), also flipper-slapping and breaching. Long-finned Pilot Whales often associate with Bottlenose Dolphins.

STATUS AND DISTRIBUTION: This species is very common in all cold temperate to sub-polar waters, with the exception of the North Pacific. In European waters, they are regularly seen inshore and offshore between Iceland and North Africa. Because the ranges of Long-finned and Short-finned Pilot Whales overlap in the Bay of Biscay, and identification is very difficult at sea, observations should generally be recorded simply as 'pilot whale'. However, as Short-finned Pilot Whales are considered to be extremely rare in the Bay of Biscay, it is likely that vast majority of sightings are of Long-finned Pilot Whale. Most sightings in the European Atlantic occur in deep water, but seasonal inshore movements take place sporadically into coastal waters to hunt migrating prey such as cuttlefish, squid and fish.

SURFACING SEQUENCE

→ DIRECTION OF TRAVEL

The distinctive head appears first …

… followed by the dorsal fin.

The long back and tail-stock show next, revealing the forward positioning of the fin.

The long back and tail-stock arch as the animal dives.

The dorsal fin of the male (*left*) is distinctive, immature and female individuals (*right*) have more dolphin-like fins.

SIMILAR SPECIES: The only reliable way of distinguishing Long-finned Pilot Whale from the much rarer Short-finned Pilot Whale (*page 74*) is on the length of the pectoral flippers. Flipper length varies in both species but averages 16% of body length in Short-finned compared with 22% in Long-finned Pilot Whale. Although False Killer Whale (*page 75*) can look similar to female pilot whale, False Killer Whale has a centrally-placed fin, but a view of the distinctive head shape of either species will confirm identification.

IDENTIFICATION: The Long-finned Pilot Whale is easily identified from all but the Short-finned Pilot Whale by its distinctive shape and behaviour. Coloration is a glossy black, with a narrow white underbelly. On surfacing, the bulbous, rounded head and distinct bushy blow precedes a robust body with the dorsal fin set well forward on the back.

The relative flipper lengths of Long-finned Pilot Whale (TOP) and Short-finned Pilot Whale (BOTTOM).

The different head shapes, but also the position of the dorsal fin should help distinguish Long-finned Pilot Whale (LEFT) from False Killer Whale (RIGHT).

A female Long-finned Pilot Whale dives.

Behind the dorsal fin, the back and tail-stock are long.

Fin positions of Bottlenose Dolphin (TOP) and Long-finned Pilot Whale (BOTTOM).

The dorsal fin is distinctly broad-based, a feature which is most striking in mature males. Mature males also develop broad, muscular shoulders as they age. Immature animals have pointed, dolphin-like dorsal fins, which, being positioned relatively close to the head, should distinguish them from most other dolphin species.

A group of Long-finned Pilot Whales logging.

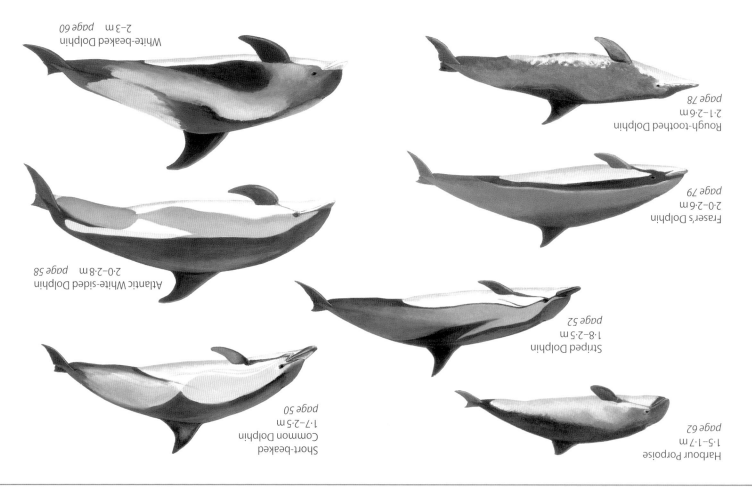

Short-beaked Common Dolphin

Delphinus delphis

F: Dauphin commun
ESP: Delfín común

Adult length:	1.7–2.5 m
Group size:	Generally 1–500
Behaviour:	Highly active, fast swimmer capable of impressive acrobatics

STATUS AND DISTRIBUTION:

Short-beaked Common Dolphin has a worldwide distribution in tropical, sub-tropical and temperate zones, and is found both in shelf waters and beyond the edge of the continental shelf. It is the most abundant cetacean in the European Atlantic and appears to be widespread. Unsurprisingly, it commonly strands on the Atlantic coasts of France, Spain and south-west England. The number of strandings has increased alarmingly over the last decade, with many dolphins showing injuries consistent with accidental capture in trawler nets.

BEHAVIOUR:

Short-beaked Common Dolphins are fast, energetic swimmers and are readily attracted to the bow-wave or wake of passing vessels. They are frequently observed porpoising clear of the water. Elsewhere in the world, this species is occasionally seen in huge pods numbering thousands of animals, but in the European Atlantic the average pod size is 20, with the largest pods containing about 500. Short-beaked Common Dolphins sometimes associate with other species, principally with Striped Dolphins, but occasionally with Bottlenose Dolphins and pilot whales.

Large groups of travelling Short-beaked Common Dolphins are typically made up of a number of smaller pods of animals, moving in unison. Lots of low leaps and a considerable amount of splashing are characteristic. At a distance the 'clumped' nature of the group and consistently low height of the splashes is a good indicator of this species.

J F M A M J J A S O N D

SURFACING SEQUENCE

DIRECTION OF TRAVEL →

IDENTIFICATION: A streamlined dolphin with a long, slender, dark beak and a tall, falcate dorsal fin. The main identification feature is a characteristic, 'figure-of-eight' pattern on the flanks. At close range, the anterior half of the 'figure-of-eight' appears tawny-yellow or brown and the posterior half grey. Close views also reveal a thin black stripe extending from the flipper to the beak and a further dark stripe running from the eye to the beak.

SIMILAR SPECIES: If seen distantly or in poor weather conditions may be mistaken for Striped Dolphin (*page 52*), due to its similar size and shape. However, the distinctive flank pattern is the best way of distinguishing Short-beaked Common Dolphin from all other species.

The dark colour of the back extends down into the central part of the 'figure-of-eight' to form a distinctive 'point' directly below the dorsal fin.

The central part of the dorsal fin is often pale.

From above the tawny-yellow pattern is still apparent.

Striped Dolphin

Stenella coeruleoalba

F: Dauphin bleu et blanc
ESP: Delfín listado

Adult length:	1.8-2.5 m
Group size:	Generally 10-500
Behaviour:	Perhaps even more acrobatic than Common Dolphin, often cautious around boats, swims in tightly-packed pod

BEHAVIOUR: Striped Dolphin is a highly acrobatic species, often seen leaping clear of the water. In the European Atlantic, pods average 25 individuals, but can number as many as 200. This species is regularly seen in association with feeding Fin Whales, sometimes riding the bow-wave created as the whales surface. Striped Dolphins are often cautious when in the vicinity of vessels, usually moving slowly in a tightly-packed pod, before speeding up and leaping high in the wake once a vessel has passed.

A travelling pod of Striped Dolphins is usually very active, tightly packed and with some individuals breaching, somersaulting and leaping up to 7 m above the surface. At a distance, the compact nature of the group, and variable height of the splashes, is a good indicator of this species.

STATUS AND DISTRIBUTION: Striped Dolphin has a worldwide distribution, but is confined to tropical, sub-tropical and warm temperate waters, generally preferring deep water beyond the continental shelf edge. This is one of the most abundant dolphin species in the European Atlantic and is regularly recorded in offshore waters south of 47°N. It rarely strays into shallow water and there have only been a very small number of animals recorded from the coastal waters of north-west France, south-west England and Ireland. However, this species regularly strands further south along the Atlantic coasts of France and Spain.

Striped Dolphins often form mixed pods with Short-beaked Common Dolphins and, because the latter species loves to bow-ride, Striped Dolphins may be observed doing the same.

J F M A M J J A S O N D

52

IDENTIFICATION: The Striped Dolphin is a slender species with a distinctive flank pattern. It has a gently-sloping forehead, dark, prominent beak and a slightly hooked, triangular dorsal fin. Similar in size and shape to Short-beaked Common Dolphin, the underparts from the throat to the base of the tail-stock are white or pinkish.

SIMILAR SPECIES: In poor light, or at long range, Striped Dolphin can be quite difficult to distinguish from Short-beaked Common Dolphin (*page 50*). Both species are broadly similar in size and shape and cannot be separated unless the distinctive flank patterning is seen. Confusion is also possible with the extremely rare Fraser's Dolphin (*page 79*). However, Fraser's Dolphin has a smaller dorsal fin and a shorter beak, with some individuals also showing a much thicker flank stripe.

A thin, dark stripe runs along the lower flanks from the eye to the underside of the tail-stock. This stripe can be difficult to see unless the dolphin jumps clear of the surface.

The upperparts are dark grey, with a distinctive pale blaze that sweeps from the flank back and up towards the dorsal fin.

From above, although the blaze may not be obvious, Striped Dolphin has a distinctive appearance.

Bottlenose Dolphin

Tursiops truncatus

F: Grand dauphin
ESP: Delfín de dientes rugosos

Adult length:	1.9–3.9 m
Group size:	1–50
Behaviour:	Highly active, capable of great speed and amazing acrobatics

BEHAVIOUR: The Bottlenose Dolphin is most commonly seen in groups of between five and 50 individuals, which, on occasion, join other dolphin groups, including other species, to form larger pods numbering several hundred animals. These large congregations regularly include pilot whales. Bottlenose Dolphins are very active and curious animals and are capable of travelling at great speed. They regularly ride in the bow-wave or wake of passing ships and are capable of amazing acrobatics.

STATUS AND DISTRIBUTION: This cosmopolitan species occurs throughout temperate and tropical seas. In most areas, including the European Atlantic, there seem to be two population types: nearshore and offshore, which

probably never interbreed. The nearshore type can be found in bays, lagoons and estuaries, with important populations present in the Shannon Estuary, Ireland; Cardigan Bay, Wales; Cornwall, England; the Channel Islands; Brittany and Normandy, France; and along the north coast of Spain. The offshore type ranges more widely in shelf waters and pelagic waters beyond the continental slope, and is known to feed regularly along the shelf-edge on fish such as Blue Whiting and Hake.

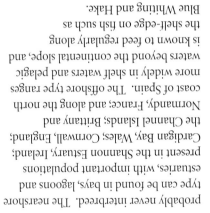

54

IDENTIFICATION: A large, robust dolphin which, due to its widespread use in aquaria and popularity with the media, is one of the most familiar of all the cetaceans.

SIMILAR SPECIES: The large size, plain grey coloration, and lack of distinct flank markings distinguish the Bottlenose Dolphin from most other dolphins. Confusion is possible with Rough-toothed Dolphin (*page 78*) and the Harbour Porpoise (*page 62*). However, the distinct bulging forehead and short, stubby beak should help to separate it from the former, and the much larger size and falcate dorsal fin from the latter.

The body coloration is uniformly grey above, fading to paler grey flanks and a lighter belly.

The curved forehead leads to a distinctly short and stubby beak.

The dorsal fin is centrally-placed, tall and falcate.

Risso's Dolphin

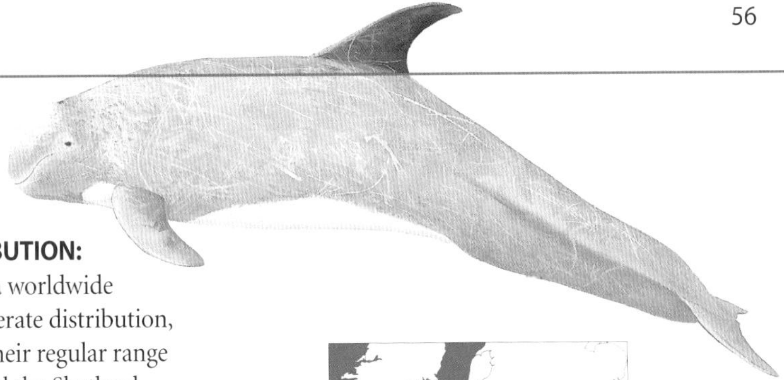

Grampus griseus

F: Dauphin de Risso
ESP: Delfin gris

Adult length:	2.6–3.8 m
Group size:	1–15
Behaviour:	Can be energetic, spy-hopping, lob-tailing and breaching, but more frequently engages in slow travel or logging

BEHAVIOUR: Risso's Dolphins feed mainly on fish and squid at night, and spend much of the day engaged in slow travel or logging at the surface. They are highly inquisitive animals and often breach, spy-hop, and tail-slap. Most sightings in the European Atlantic involve animals milling leisurely at the surface, or travelling slowly. Pod sizes generally comprise less than 15 individuals.

STATUS AND DISTRIBUTION:

Risso's Dolphins have a worldwide tropical to warm temperate distribution, the northern limit of their regular range in Europe being around the Shetland Islands. In the Bay of Biscay, sightings have been made in all regions and water depths, but encounters are infrequent. Further north, Risso's Dolphins are seen regularly in the western English Channel and southern Irish Sea, including off Cornwall, England, and Pembrokeshire, Wales. They are also sighted regularly close to shore off south-west Ireland.

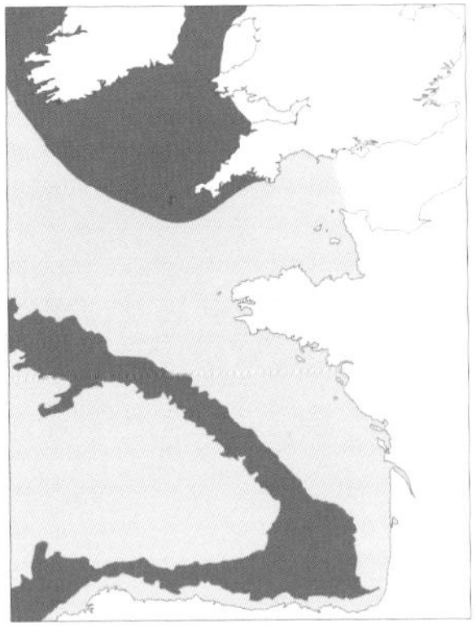

J F M A M J J A S O N D

SURFACING SEQUENCE

A slowly-travelling animal (the blow may be visible at close range in calm weather) will surface leisurely, head first ...

... followed by the back and dorsal fin ...

... after which the animal assumes its characteristic profile of head, back and swept-back dorsal fin ...

... before rolling over slowly exposing the fin and tail-stock (occasionally fluking).

→ DIRECTION OF TRAVEL

SIMILAR SPECIES: Views of the bulbous forehead, with a lack of a distinctive beak, white coloration and extensive scarring rule out all other dolphins in the region. Surprisingly, perhaps, the most likely confusion species in good light is Cuvier's Beaked Whale (*page 38*), mature individuals of which can also appear extensively pale coloured and scarred on the upper surface. However, the tall, centrally-placed dorsal fin and bulbous head-shape of Risso's Dolphin is clearly different. Risso's Dolphins are also similar in shape to several members of the blackfish family, particularly pilot whales (*pages 46 & 74*), with which they often associate. Several of the above features should be noted to distinguish them in poor light.

A surfacing Adult Risso's Dolphin showing the distinct falcate dorsal fin.

A close encounter with this immature animal reveals the distinctive bulbous head with its short, indistinct beak.

IDENTIFICATION: The distinctive shape and coloration of Risso's Dolphin means that it is unlikely to be mistaken at close range. Initial impressions are of a large, robust dolphin with a tall, falcate dorsal fin.

The body is uniform dark grey to white in coloration depending on maturity. The dorsal fin is centrally-placed, tall and falcate. Adults become paler and more heavily scarred as they get older, although the dorsal fin and adjacent back usually remain distinctly darker. The often extensive scarring is the result of battles with other Risso's Dolphins.

Atlantic White-sided Dolphin

Lagenorhynchus acutus

F: Dauphin à flancs blanc
ESP: Delfín de flancos blancos

Adult length:	2.0–2.8 m
Group size:	1–50
Behaviour:	Acrobatic and energetic, breaches, and lob-tails

BEHAVIOUR: A very fast moving dolphin, Atlantic White-sided Dolphins often travel rapidly at the surface, revealing only their upper backs as they power through the water. At other times they are very acrobatic and energetic, often exhibiting exuberant breaching and tail-slapping displays. They are regularly encountered in pods numbering up to 50 animals, although larger groups of several hundred individuals are not uncommon.

STATUS AND DISTRIBUTION: The Atlantic White-sided Dolphin is confined to the cool temperate and sub-arctic waters of the North Atlantic.

In Europe, it is regularly recorded during the summer months in offshore waters to the west of Britain and Ireland, north to the Faroe Islands, Iceland and western Norway. This is one of the scarcest species of dolphin to occur in the European Atlantic, with the waters off northern Spain representing the southernmost extent of its range. They have only been recorded on a handful of occasions from ferries in the Bay of Biscay, and only a very small number of animals are sighted or strand along the Atlantic coasts of England, Wales, France, and Spain each year. In Ireland, this species is only rarely seen close to shore, but is known to be fairly abundant along the edge of the continental shelf to the south and west.

SIMILAR SPECIES: At long range, the Atlantic White-sided Dolphin could be confused with several other dolphins, although the yellow or tan flank stripe is diagnostic. The most likely confusion species is the similarly-sized White-beaked Dolphin (*page 60*), which also has a prominent, falcate dorsal fin. However, unlike the Atlantic White-sided Dolphin, this species has white or pale dorsal markings below and behind the dorsal fin.

IDENTIFICATION: A large, robust dolphin with a markedly thick tail-stock. It has a tall, sickle-shaped dorsal fin situated halfway along the back. The short beak, back and dorsal fin are dark-coloured, the flanks are pale grey, and the belly white.

The principal identification feature is the sharply-defined white patch on the flanks which extends along the tail-stock as an elongated yellow stripe.

White-beaked Dolphin

Lagenorhynchus albirostris

F: Dauphin à bec blanc
ESP: Delfín de hocico blanco

BEHAVIOUR: The White-beaked Dolphin is a fast and powerful swimmer, which can be very active and has regularly been observed breaching and tail-slapping. Like many other dolphins, it is not shy and will ride in the bow-waves or wakes of ships, boats or even large whales. Rarely seen alone, White-beaked Dolphins tend to occur in small pods, with up to 20 individuals being recorded together.

STATUS AND DISTRIBUTION: The White-beaked Dolphin is confined to the cool and sub-arctic waters of the North Atlantic. In Europe, it is most commonly found around the coasts of northern Britain, north to Iceland. The coastal waters off northern Spain represent the southernmost limit of this species' range. In the European Atlantic White-beaked Dolphins have only been recorded on a few occasions over the shelf and shelf-edge of the Bay of Biscay. They are also only rarely sighted in the English Channel, and off Welsh and Irish coasts.

J F M A M J J A S O N D

SIMILAR SPECIES: At a distance, the White-beaked Dolphin could be mistaken for several other dolphins, but confusion is most likely with its close relative the Atlantic White-sided Dolphin (*page 58*). However, the latter species can be distinguished by the long, discrete, oval-shaped white patch on the flanks, which extends along the tail-stock as a yellow band. In addition, Atlantic White-sided Dolphins do not have a pale 'saddle' behind the dorsal fin.

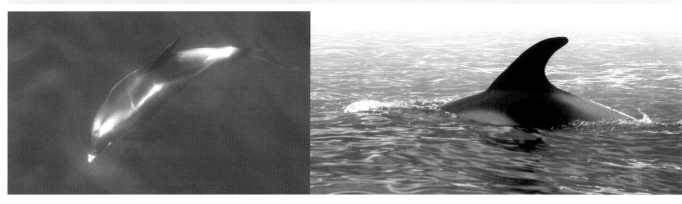

IDENTIFICATION: A large and very robust dolphin with a short, thick rounded beak and thick tail-stock. The belly is white as far back as the tail-stock, the dorsal fin is tall and sickle-shaped, and the beak is usually pale grey or white.

One of the most useful identification features is the white or pale grey markings on the flanks. The most distinctive of these markings is a pale grey area which extends from the upper flanks in front of the dorsal fin towards the tail-stock, and onto the back behind the dorsal fin, giving the impression of a pale 'saddle'.

Harbour Porpoise

Phocoena phocoena

F: Marsoin commun
ESP: Marsopa

Adult length:	1.5–1.7 m
Group size:	1–6
Behaviour:	Slow swimmer, unobtrusive, shy, usually moves away from boats

BEHAVIOUR: Harbour Porpoises are usually sighted alone or in small groups, although larger aggregations are regularly observed in certain locations. They generally travel quite slowly, are unobtrusive and very difficult to observe in rough weather. Unlike many dolphin species, Harbour Porpoises are shy and rarely approach boats; indeed they often move away from approaching vessels.

A surfacing Harbour Porpoise can be challenging to spot and observe, being highly unobtrusive in all but the calmest conditions.

STATUS AND DISTRIBUTION: The Harbour Porpoise is restricted to shallow water over the continental shelf and often favours coastal areas. It is widespread in the cold temperate and sub-arctic waters of the Northern Hemisphere. In the eastern North Atlantic, it ranges from Iceland south to the coast of Senegal, including the North and Baltic Seas and the western Mediterranean. This species is recorded on a regular basis around the Brittany Coast, in the western English Channel and along the Welsh and Irish coasts, with late summer congregations sometimes reaching 50–100 feeding animals off some headlands. Overall, the European population has declined significantly in the last 30 years due to increased levels of pollution, loss of prey through over-fishing and accidental capture in fishing nets.

J F M A M J J A S O N D

SURFACING SEQUENCE

→ DIRECTION OF TRAVEL

The top of the head and back as far back as the dorsal fin appear first in a very shallow roll ...

... the head then submerges ...

... and the animal submerges, the fin being the last part to remain above the surface.

SIMILAR SPECIES: At long range, this species may be confused with Bottlenose Dolphin (*page 54*), which also shows a centrally-placed dorsal fin and uniform grey upperparts. However, when seen well, the diminutive size, distinctive shape of the dorsal fin, and unobtrusive behaviour should distinguish the Harbour Porpoise from all other cetaceans in the region.

IDENTIFICATION:
The Harbour Porpoise is the smallest cetacean occurring in the European Atlantic. It has a robust body, a small, rounded head and no beak.

Fin shape of Bottlenose Dolphin (LEFT) and Harbour Porpoise (RIGHT).

The upperparts of the body are dark grey, merging to lighter grey on the flanks; the underside of the body is white. Calves and juveniles often have brownish backs.

The most obvious identification feature is the centrally-placed dorsal fin which is small, low and triangular in shape.

RARE SPECIES

The European Atlantic is home to an outstanding diversity of cetaceans, with around one-third of the world's species having been recorded in the region. This high diversity is a reflection of the occasional occurrence of a number of species which are normally found in other parts of the North Atlantic. The occurrence of these 'rarities' adds to the excitement of whale-watching in the European Atlantic, and is most likely to be due to the temperate geographical location of the region and the number of distinct circulatory current systems that are in operation here.

Some of these currents bring cool water from the north, whereas others carry warm water from the south. The relative strength of these currents does, however, vary from year to year. Consequently, species with high-latitude summer distributions, such as Sei Whale, Blue Whale, Northern Bottlenose Whale, White-beaked Dolphin and Atlantic White-sided Dolphin, are occasionally recorded. Conversely, years in which warmer waters become more prevalent

may be responsible for the appearance of tropical species such as Blainville's Beaked Whale and Melon-headed Whale, both of which have stranded along the French coast in recent years.

One oceanic current that may influence the presence of rarer species in the Bay of Biscay runs northwards along the Portuguese continental slope and into the southern Bay of Biscay between October and March, pushing warm water along the coast of northern Spain. This is due to a relaxation of southerly winds that are persistent off North Africa and Portugal during the summer months.

It may be no coincidence that this coastline has received winter strandings of both Short-finned Pilot Whales and False Killer Whales – two species that are generally distributed well to the south of mainland Europe.

By far the most influential current in the European Atlantic is the Gulf Stream. Originating near the Bahamas, it travels north-eastwards, passing to the west of Ireland and continuing on

past Scotland. Its warm waters have a significant affect on the climate of the entire western European landmass. This current also has a strong affect on the distribution of cetaceans, with a number of warm-water species occurring at higher latitudes than in other parts of the world.

This section includes a number of species that we have defined as 'rarely seen' in the European Atlantic. Many of these species share a number of identification features with the regularly occurring cetaceans in the first half of this book.

The first step in identifying a rarity must therefore always be to review the possibility that your sighting is simply an unusual view of, or example of, a commoner species. Indeed, several of the species in this section are so rare that photographs or video footage are essential to confirm identification with the relevant whale and dolphin research organisations (see page 83). Encountering rare cetaceans is always a thrill, and the more time you spend at sea, the greater your chances of encountering something that you really did not expect!

Beluga

Delphinapterus leucas

F: Béluga
ESP: Beluga

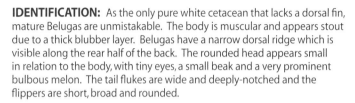

BEHAVIOUR: Belugas hunt for fish, squid and crustaceans on the seabed, diving for between two and 15 minutes and reaching depths of up to 650 m.
At the surface they often appear sluggish, regularly forming chorus-lines of four to six animals.
Groups typically involve two to 25 whales, often all of the same age and sex, although summer gatherings in selected estuaries may number in the thousands. Nicknamed 'Sea Canaries', Belugas are exceptionally vocal, with a range of clicks, squeaks, squawks and whistles (likened to the string section of an orchestra playing out of tune!) that are audible even above water.

STATUS AND DISTRIBUTION: Restricted to Arctic and sub-Arctic waters, Belugas spend the winter and spring close to the pack ice or underneath it, finding patches of open water to use as breathing holes. Ice melt during early summer frees up coastal bays, estuaries and inlets which are favoured by the whales until the ice returns in the autumn and the animals retreat once more. On rare occasions, Belugas stray into northern Europe and there has been one record in the European Atlantic. This involved a young female measuring 3·25 m which was caught in a salmon net on the River Loire approximately 18 km downstream from Nantes.

SIMILAR SPECIES: Risso's Dolphin (*page 56*) and Cuvier's Beaked Whale (*page 38*) can appear white and 'Beluga-like' at sea, but both have prominent dorsal fins.

Adult length:	4–5 m
Group size:	Social, but usually seen alone south of Arctic Circle
Breaching:	No
Blow:	Small and inconspicuous

IDENTIFICATION: As the only pure white cetacean that lacks a dorsal fin, mature Belugas are unmistakable. The body is muscular and appears stout due to a thick blubber layer. Belugas have a narrow dorsal ridge which is visible along the rear half of the back. The rounded head appears small in relation to the body, with tiny eyes, a small beak and a very prominent bulbous melon. The tail flukes are wide and deeply-notched and the flippers are short, broad and rounded.

Belugas are born uniformly dark but become paler as they mature. By the age of nine (males) and seven (females) they appear pure white, although the dorsal ridge and the edges of the flippers and flukes may remain dark. Heavy scarring may be indicative of a Polar Bear attack.

One unusual characteristic of the Beluga is its distinctive narrow neck, the vertebrae of which, unlike those of most other cetaceans, are not fused. This allows the whale to move its head from side to side and nod like humans do, and may be an adaptation for seeking out prey in the confined spaces of an ice-bound environment.

Blue Whale

Balaenoptera musculus

F: Rorqual bleu
ESP: Ballena azul

Adult length:	24–30 m
Group size:	1–2, sometimes more
Breaching:	Only young known to breach, usually at 45° angle
Deep dive:	Sometimes raises tail flukes
Blow:	Largest of all: an enormous vertical column up to 10 m tall

BEHAVIOUR: Blue Whales are usually found singly or in pairs, although larger aggregations are found towards the poles where there is a rich source of food. They occasionally show their flukes during deep dives, but only young animals have been observed breaching. They feed almost exclusively on krill.

STATUS AND DISTRIBUTION:

Blue Whale has a worldwide distribution, with summer feeding grounds in cold sub-polar waters where krill is particularly abundant. However, whaling has severely depleted the population throughout its range. After a summer feasting, the whales migrate to lower latitudes in the winter, when breeding occurs. In the North Atlantic, small numbers are regularly recorded in the Gulf of St. Lawrence, Canada, and around the coasts of Iceland and Greenland. It is an extremely rare species in the European Atlantic. Sightings of Blue Whales from ferries occur on average around once a year, and usually coincide with the peak arrival of Fin Whales during late summer.

There is also acoustic evidence to suggest that there is a wintering population of over 50 animals in waters south-west of Ireland, but these animals remain unobserved.

J F M A M J J A S O N D

SURFACING SEQUENCE

DIRECTION OF TRAVEL →

The huge blow appears first …

… then, slowly, the first part of the back. Being so large an animal the dorsal fin is still below surface.

Eventually the small dorsal fin appears …

… before the animal rolls revealing most of the tail-stock.

A travelling whale then rolls over and repeats the sequence …

… unless it goes into a deep dive when the short tail-stock and sometimes the fluke appear.

SIMILAR SPECIES: At long range, or if poorly seen, this species can be confused with Fin Whale (*page 28*). Both are very large, with a tall, vertical blow and uniform greyish back. At relatively close range, Fin Whales appear darker, often with pale chevrons over the back, but without any pale mottling. Fin Whales also have a much more prominent dorsal fin which appears more rapidly after the blowhole on surfacing.

The body is streamlined, with a broad head, rounded rostrum and large splashguard.

IDENTIFICATION: The Blue Whale is the largest animal ever to have lived on Earth, and is most easily identified by its enormous blow, prominent splashguard, mottled body pattern and tiny dorsal fin.

The body colour is dark bluish-grey, mottled with light grey blotches.

Beyond the small dorsal fin, the tail-stock is short and very robust.

Blue Whales sometimes fluke at the start of a deep dive.

The massive blow can reach 10 m in height.

The dorsal fin is very small relative to body size and is situated far along the back.

North Atlantic Right Whale

Eubalaena glacialis

F: Baleine franche boréale
ESP: Ballena franca

Adult length:	11–18 m
Group size:	1–3, sometimes more
Breaching:	Often, sometimes repeatedly
Deep dive:	Usually raises tail flukes
Blow:	Distinctive, high, 'V'-shaped blow when viewed from front or rear

BEHAVIOUR: North Atlantic Right Whales are slow swimmers and are inquisitive and approachable. They can exhibit a range of behaviours such as lob-tailing, flipper-waving and breaching, sometimes up to ten times in a row, landing with a resounding splash. North Atlantic Right Whales subsist almost entirely on tiny copepods (small crustaceans).

STATUS AND DISTRIBUTION:
Restricted to the North Atlantic, the North Atlantic Right Whale may now be the most endangered whale on earth, with a population estimated to be just 350 animals. It is present across a narrow band of the eastern seaboard of North America between Canada and Florida. The European population was once widespread from Spain to Norway and Iceland in the summer, probably retreating to breeding grounds off West Africa for the winter months. This species was once abundant in the Bay of Biscay, and early taxonomists named it *Eubalaena biscayensis*. Sadly, it is now generally considered extinct in European seas due to extensive whaling, which began as early as the 11th century with the Spanish Basque fishery. As a result there has been only one sighting in the European Atlantic in recent times: on 5th December 1993 a lone whale was seen swimming and breaching 200 m off the coast of Corunha, Spain.

SURFACING SEQUENCE

DIRECTION OF TRAVEL →

The distinctively calloused head and blow ('V'-shaped unless seen from the side) appear together first ...

... followed by the back, with the top of the head and often part of the jawline still visible.

The animal then rolls, and the lack of a dorsal fin becomes apparent.

A travelling whale then rolls over and repeats the sequence ...

... unless diving when the flukes appear, the angle of the tail-stock and flukes indicate the depth of the dive.

IDENTIFICATION: The North Atlantic Right Whale is strikingly different from the other large whales of the European Atlantic due to its 'V'-shaped blow, broad, flat and very dark back lacking a dorsal fin and characteristic pale lumps on the top of the head.

SIMILAR SPECIES: Confusion is possible with other large whales which have deceptively variable blows and may sometimes surface so that their dorsal fin remains below the surface.

The huge head has light-coloured lumps called callosities, particularly around the rostrum and blowholes, above the eyes, and along the lower jaw. Callosities are hardened patches of skin which become pale due to infestations of whale lice.

The tail flukes, which are often raised on sounding, are very large and all black, with smooth edges, pointed tips and a deep central notch.

The first clue to a right whale's presence is usually its distinctive 'V'-shaped blow.

The body is large and bulky and the animal often remains quite low in the water.

Pygmy Sperm Whale

Kogia breviceps

F: Cachalot pygmée
ESP: Cachalote pigmeo

Adult length:	2.7-3.7 m
Group size:	1-6
Breaching:	Leaps vertically
Deep dive:	Tail flukes not raised
Blow:	Blow low and faint; blowhole offset slightly to the left

BEHAVIOUR: Pygmy Sperm Whales are typically seen moving slowly or hanging motionless at the surface. On diving, the body tends to sink below the surface, rather than rolling forwards. When disturbed, this whale sometimes emits a reddish-brown anal liquid to try and disorientate as a defence mechanism. Vertical breaches have been recorded.

STATUS AND DISTRIBUTION: Most information on the global distribution of Pygmy Sperm Whale comes from strandings, which indicate that it occurs in temperate to tropical deep waters. Both Pygmy and Dwarf Sperm Whales are often described as rare throughout their range, although the paucity of sightings may be due in part to their shy behaviour, small size and generally pelagic distribution. Seven Pygmy Sperm Whales stranded along the French Biscay coast and five stranded on the Galician coast of north-west Spain between 1984 and 2001. More recently, in January 2002, a female washed up dead in Devon, south-west England and in July 2004, one stranded east of Bilbao, Spain.

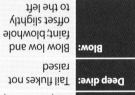

J F M A M J J A S O N D

Based on these few, widely distributed strandings, it seems likely that this species occurs regularly in the deeper waters of the Bay of Biscay. There is now also evidence of its occurrence at sea. In 1992 and 1993 researchers travelling with French albacore tuna fishermen to the west of the Bay of Biscay recorded an entangled Pygmy Sperm Whale and in August 2004, a lone animal, probably of this species, was seen close to a ferry by experienced observers.

IDENTIFICATION: The diminutive size of the Pygmy Sperm Whale means that it is most likely to be mistaken for a dolphin or porpoise. However, in structure and behaviour it is quite unlike either. Pygmy Sperm Whales are greyish in colour with a blunt, squarish head and a small, falcate dorsal fin.

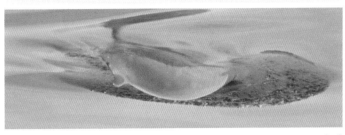

SIMILAR SPECIES: The distinctive body shape and lethargic behaviour help to distinguish Pygmy Sperm Whale from similarly-sized dolphin species with relative ease. However, its similarity to Dwarf Sperm Whale (*below*) means that many good sightings at sea fail to distinguish between the two species. Pygmy Sperm Whale differs from the slightly smaller Dwarf Sperm Whale by its smaller dorsal fin positioned further than half way along the back.

Dwarf Sperm Whale
Kogia simus

F: Cachalot nain
ESP: Cachalote enaño

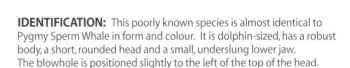

BEHAVIOUR: Dwarf Sperm Whales occasionally breach, but they are more likely to be seen travelling slowly or resting at the surface. When disturbed they sometimes emit a jet of reddish black ink as a defence mechanism.

Adult length:	2·1–2·8 m
Group size:	1–10
Breaching:	Leaps vertically
Deep dive:	Tail flukes not raised
Blow:	Blow low and faint; blowhole offset slightly to the left

IDENTIFICATION: This poorly known species is almost identical to Pygmy Sperm Whale in form and colour. It is dolphin-sized, has a robust body, a short, rounded head and a small, underslung lower jaw. The blowhole is positioned slightly to the left of the top of the head.

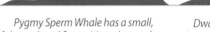

Pygmy Sperm Whale has a small, falcate, dorsal fin positioned more than halfway along the back.

Dwarf Sperm Whale has a broad-based dorsal fin that is positioned centrally on the back.

STATUS AND DISTRIBUTION: Records of Dwarf Sperm Whale suggest a global distribution in deep-water temperate to tropical seas. Given that this species is known from only two strandings in the European Atlantic, it is likely to be rare. The first was a male in very shallow water near Port Louis on the southern coast of Brittany, France, on 24th October 1991. An attempt was made to return it to open water but it failed. The second, on 15th November 1999, involved a young, 1·75 m long, female in the Gironde estuary, France (45°40'N). There is also one sighting in the European Atlantic thought to be possibly of this species. On 20th August 2001, five *Kogia* sperm whales were observed logging and swimming slowly in waters 4,000 m deep in the Bay of Biscay. Their dorsal fins were larger than those typical for Pygmy Sperm Whale.

J F M A M J J A S O N D

SIMILAR SPECIES: This small cetacean is superficially similar in size and shape to several species of dolphin but the blunt forehead lacking a beak, and generally lethargic behaviour are distinctive. Separation from Pygmy Sperm Whale (*above*) is much more challenging, but Dwarf Sperm Whale is slightly smaller in size with a larger dorsal fin, which is broader at the base, and positioned centrally on the back.

Gervais' Beaked Whale

Mesoplodon europaeus

F: Mésoplodon de Gervais
ESP: Zifio de Gervais

BEHAVIOUR: The only observations of Gervais' Beaked Whale in recent times have come from the Azores and Canary Islands. Due to uncertainty about the identity of most sightings, details of behaviour have yet to be confirmed. The lack of sightings, however, suggest that this species is probably unobtrusive and shy.

STATUS AND DISTRIBUTION: Gervais' Beaked Whale is almost completely unknown at sea. In Europe it has been photographed and identified alive on only one occasion – close to the Canary Islands. Most of our limited knowledge of this species therefore comes from a handful of strandings. The species' stronghold appears to be in the western North Atlantic and there are few European records, including the first known specimen, which was found floating in the English Channel in 1840. Its occurrence in the Bay of Biscay was confirmed when a 4·09 m long male stranded at Biscarrosse, France (44°20'N) in 1999. There was also a single stranding in County Sligo, Ireland, in 1989.

SIMILAR SPECIES: Similar in size, shape and coloration to other beaked whales. Beak length is generally considered to be intermediate between the stubby beak of True's Beaked Whale (*page 43*) and the long, slender beak of Sowerby's Beaked Whale (*page 42*).

Adult length:	4·6–6 m
Group size:	Unknown
Breaching:	Unknown
Blow:	Unknown

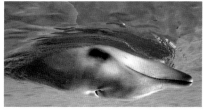

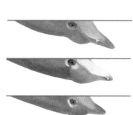

The beak length of Gervais' Beaked Whale (middle) is intermediate between True's Beaked Whale (top) and Sowerby's Beaked Whales (bottom).

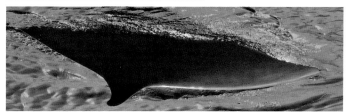

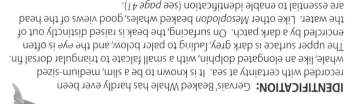

IDENTIFICATION: Gervais' Beaked Whale has hardly ever been recorded with certainty at sea. It is known to be a slim, medium-sized whale, like an elongated dolphin, with a small falcate to triangular dorsal fin. The upper surface is dark grey, fading to paler below, and the eye is often encircled by a dark patch. On surfacing, the beak is raised distinctly out of the water. Like other *Mesoplodon* beaked whales, good views of the head are essential to enable identification (see *page 41*).

The small forehead curves gently down to a shortish, slender beak. In adult males, a pair of teeth emerge along the lower jaw one-third of the way from the tip: this is the only known diagnostic identification feature for the species.

Blainville's Beaked Whale

Mesoplodon densirostris

F: Mésoplodon de Blainville
ESP: Zifio de Blainville

BEHAVIOUR: Blainville's Beaked Whales are believed to be unobtrusive in their behaviour, although they are known to breach.

Adult length:	4–6 m
Group size:	1–6
Breaching:	Unknown
Blow:	Small blow projects forwards

STATUS AND DISTRIBUTION:
Blainville's Beaked Whale is perhaps the most widespread *Mesoplodon* species, occuring throughout the world's oceans except the Arctic, with the great majority of records coming from the tropical Atlantic north to the Canary Islands. Within the European Atlantic, it is likely to be extremely rare. In recent years, a stranding in Abaraeron, Wales on 18th July 1993 was followed by the stranding of a 4·35 m individual at Tarnos, France on 8th January 1998. The most recent European sighting occurred to the north of the region and involved a pregnant female that stranded at Ameland on the Dutch North Sea coast on 12th April 2005. This represented only the 8th record of Blainville's Beaked Whale for Europe (outside the Canary Islands, Azores and Madeira).

IDENTIFICATION: Blainville's Beaked Whale is similar in size to the other *Mesoplodon* beaked whales. The body is reminiscent of an elongated dolphin, with a small, falcate to triangular dorsal fin, a flattish head, and uniform grey or brown coloration on the body. Mature Blainville's Beaked Whales can only be identified at sea by the unique shape of the beak (see *below* and *page 41*) and although observation of this at sea is extremely difficult, requiring an exceptionally good encounter, the confirmation of a live sighting remains a real possibility.

The lower jaw is relatively short and distinctly arched, giving the impression, at a distance, of a bump on the forehead. In mature males, the arch is capped by two large, protruding teeth which are often encrusted in barnacles.

Female Blainville's Beaked Whale (left) has a less distinctly raised jawline than the male (right).

SIMILAR SPECIES: Blainville's Beaked Whale overlaps in coloration and size with all of the other beaked whales. However, none of these species display the diagnostic raised jawline.

Short-finned Pilot Whale

Globicephala macrorhynchus

F: Globicéphale tropical
ESP: Calderón tropical

BEHAVIOUR: The behaviour of Short-finned Pilot Whales at the surface is almost identical to that of Long-finned Pilot Whales. These whales are almost always seen in family groups travelling slowly at the surface or logging. Other behaviours such as spy-hopping and flipper-slapping occur frequently. Short-finned Pilot Whales often associate with dolphins, particularly Bottlenose Dolphin.

STATUS AND DISTRIBUTION: Most commonly distributed in tropical and sub-tropical waters between 40°N and 40°S, well south of the Bay of Biscay. However, between 1982 and 1986, ten pilot whales that stranded on the north coast of Spain were positively identified as Short-finned Pilot Whales, bringing into question the northern limit of this species' range.

J F M A M J J A S O N D

Adult length:	3.5–6.5 m
Group size:	Family groups of 2–50
Breaching:	Occasional, various angles
Blow:	Distinctly bushy

SURFACING SEQUENCE

→ DIRECTION OF TRAVEL

The distinctive head appears first ...

... followed by the dorsal fin.

The long back and tail-stock show next, revealing the forward positioning of the fin.

The long back and tail-stock arch as the animal dives.

IDENTIFICATION: Short-finned Pilot Whales are fairly distinctive at sea, except where their range overlaps with Long-finned Pilot Whales. The impression is of a large, robust, black dolphin with a bulbous head, a broad dorsal fin situated well forward on the back, and a long tail-stock. The flippers are generally less than one-fifth of the body length.

A pale 'saddle' behind the dorsal fin and various pale markings are often present on the back.

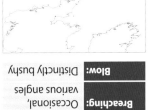

They exhibit a low, bushy blow while travelling.

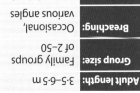

SIMILAR SPECIES: Reliable identification from Long-finned Pilot Whale (page 46) requires exceptional views of the comparatively short flippers. Some researchers suggest that the pale 'saddle' occurs far more regularly in Short-finned Pilot Whale, but this feature cannot be used alone to confirm identification. Short-finned Pilot Whale has yet to be identified positively at sea in the European Atlantic.

False Killer Whale
Pseudorca crassidens

F: Faux-orque
ESP: Orca falsa

BEHAVIOUR: Although False Killer Whales are often seen in association with other cetaceans, such as Bottlenose Dolphins, they have been recorded attacking a range of cetacean species, including Sperm Whale. Pod sizes are generally small, but several hundred have been seen together on occasion.

STATUS AND DISTRIBUTION:
False Killer Whale is a warm water species that seems to be fairly uncommon throughout its range. Although this species has been recorded as far north as the UK, the northern limit of its regular range in the Atlantic appears to be around the Straits of Gibraltar. False Killer Whales are scarce visitors to the Bay of Biscay, with only a handful of sightings in deep waters along the shelf-edge which may be linked to incursions of unusually warm water.

Adult length:	4–6 m
Group size:	2–200
Breaching:	Frequent, various angles
Blow:	Inconspicuous and bushy

J F M A M J J A S O N D

IDENTIFICATION:
False Killer Whales are large, elegant, and extremely active predators. They are typically seen in family groups, racing powerfully at the surface with their heads and upper bodies raised above the water. At other times, they travel at a more leisurely pace.

SIMILAR SPECIES: False Killer Whale is perhaps most likely confused with female or immature pilot whales (*above* & *page 46*), from which it can be distinguished by the slender, tapering head, slim body and dorsal fin placed centrally on the back, or Killer Whale (*page 44*) from which it differs in its all-dark body.

SURFACING SEQUENCE

The distinctive head appears first … … and then a forward surge brings the back and dorsal fin clear of the surface. The roll continues, revealing the central positioning of the fin … … after which the tail stock appears and the animal submerges.

← DIRECTION OF TRAVEL

Pygmy Killer Whale

Feresa attenuata

F: L'orque naine
ESP: Orca pigmea

Adult length:	2.0–2.7 m
Group size:	<50 in tropical waters
Breaching:	Occasional
Blow:	Inconspicuous, low and bushy

BEHAVIOUR: Pygmy Killer Whales tend to live in pods of 50 or less in the tropics. Although they are capable of most exuberant dolphin behaviours, they are generally not particularly acrobatic. When porpoising, both Pygmy Killer and Melon-headed Whales often raise their heads clear of the water as they move rapidly forwards in a chorus-line. They occasionally bow-ride but are often wary and shy of vessels. They regularly associate with other dolphins in tropical waters.

STATUS AND DISTRIBUTION: Pygmy Killer Whale is a tropical species that occasionally ventures into sub-tropical and warm temperate seas. Little is known about their abundance or seasonal movements. There have been a handful of reported sightings of Pygmy Killer Whales in the Bay of Biscay, notably two separate groups seen at close hand from a ferry in the Northern Bay in April 1997. However, photographs have yet to be taken that confirm the presence of this species in the region.

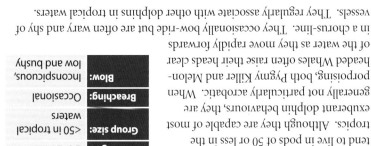

SURFACING SEQUENCE → DIRECTION OF TRAVEL

Melon-headed Whale Pygmy Killer Whale

IDENTIFICATION: Pygmy Killer Whales are stocky, predominantly black, cetaceans with a relatively bulky body. Sometimes visible is a dorsal cape that is widest below the dorsal fin, which contrasts with the paler flanks. The lips, an anchor-shaped throat patch and the hindbelly can also appear pale or white on some individuals. Identification of Pygmy Killer Whale is easiest when it is bow-riding. This allows the observer to look down on the animal and note its two most distinctive features: the shape of the head and the flippers. Pygmy Killer Whales have a rounded melon, and long flippers that are broadly rounded at the tip.

Pygmy Killer Whale

Melon-headed Whale

SIMILAR SPECIES: Pygmy Killer Whale looks very similar to Melon-headed Whale (below) and, in poor light conditions, other dolphins. Both Melon-headed Whale and Pygmy Killer Whale could also be mistaken for larger members of the blackfish family, such as young pilot whales (pages 46 & 74). However, both pilot whale species differ in having a dorsal fin positioned forward of centre on the back, a pale saddle patch and a more bulbous melon; the False Killer Whale has a more slender, tapering head-shape.

Melon-headed Whale
Peponocephala electra

F: Dauphin électre
ESP: Falsa orca de cabeza de mélon

BEHAVIOUR: Melon-headed Whales are highly gregarious, occurring in groups of up to 1,500, although vagrants to European seas may involve much smaller pods or lone individuals. Surface behaviour includes spy-hopping, tail-slapping and high leaps. When porpoising, Melon-headed Whales often raise their heads clear of the water as they move rapidly forwards in a chorus-line. They occasionally bow-ride but are often wary and shy of vessels. They regularly associate with Fraser's Dolphins and Rough-toothed Dolphins in tropical waters.

Adult length:	2·0–2·7 m
Group size:	>50 in tropical waters
Breaching:	Occasional
Blow:	Inconspicuous, low and bushy

IDENTIFICATION:
Melon-headed Whales are easily told from most dolphin species by their all-dark coloration and the lack of a protruding beak. They are fast-moving cetaceans with a relatively bulky body forward of the centrally-positioned, falcate dorsal fin, in contrast to the rear of the animal which tapers to a thin tail-stock. The colour of the skin is dark grey, and there is a dorsal cape that is widest below the dorsal fin. The flanks are paler, and there are several areas that are often pale or white, including the lips, an anchor-shaped throat patch and the hindbelly.

Melon-headed Whale has a relatively pointed (melon-shaped) head and sharply-pointed flippers, which differentiate it from the very similar Pygmy Killer Whale, though these features can be difficult to observe at sea.

STATUS AND DISTRIBUTION: An equatorial species that is rarely sighted in sub-tropical and warm temperate seas. The first European record involved a skull found near Charlestown, Cornwall, in September 1949. Then, on 27th August 2003, two Melon-headed Whales stranded alive near La Rochelle, France. Both were refloated, but one, an adult male, was discovered dead on a nearby beach two days later.

In good light, the face of Melon-headed Whale appears darker than the flanks, giving the animal a masked appearance.

Both Melon-headed and Pygmy Killer Whales have similar surfacing sequences. Although both species rest motionless at the surface, when travelling Melon-headed Whales are generally the more energetic species and hence more likely to lift their heads clear of the surface and show their tail flukes when diving.

← DIRECTION OF TRAVEL   **SURFACING SEQUENCE**

SIMILAR SPECIES: Telling Melon-headed Whale from Pygmy Killer Whale is difficult at sea, as the key features are subtle and not easily seen. In contrast to Melon-headed Whale, Pygmy Killer Whale has a rounded melon, and has long flippers that are broadly rounded at the tip.

Rough-toothed Dolphin

Steno bredanensis

F: Steno
ESP: Steno

Adult length:	2.1–2.6 m
Group size:	1–50
Behaviour:	Sometimes shows a characteristic swimming pattern described as 'skimming across the surface'

BEHAVIOUR: The Rough-toothed Dolphin is a relatively poorly-known species which has been seen most frequently in groups of between 10 and 20, although pods of over 100 have been reported. It does not appear to be a particularly active species, but will often ride the bow-wave of ships. Occasionally when swimming at high speed, Rough-toothed Dolphins hold their heads and chins above the surface of the water in a characteristic behaviour described as 'skimming'.

STATUS AND DISTRIBUTION: Rough-toothed Dolphins have a worldwide distribution in tropical, subtropical and warm temperate waters. They are normally found in deep waters beyond the edge of the continental shelf. In Europe, this species is regularly seen around the Canary Islands and occasionally in the Azores and Madeira.

There are no documented sightings at sea in the European Atlantic, but two strandings have occurred on the Atlantic coast of France. This species appears to be an extreme rarity in the Bay of Biscay and seems most likely to occur during warm water years.

SIMILAR SPECIES: The wholly dark grey and robust body makes this species appear very similar to Bottlenose Dolphin (page 54). However, when seen well, the distinctive head structure and dark cape should distinguish Rough-toothed Dolphin from the latter species.

IDENTIFICATION: A fairly robust dolphin. The dorsal fin is wide at the base, and pointed but not hooked. The upper-back is dark grey, forming a cape, becoming lighter towards the flanks which often have white or pinkish blotches. The belly and undersides of the tail-stock are usually white. The beak may also show whitish or pinkish patches, particularly around the lips.

The head-shape is distinctive, being conical with a low melon and long, slender beak. Unlike other long-beaked dolphins, there is no distinct differentiation between the beak and the forehead, giving this species a slightly reptilian appearance.

Fraser's Dolphin

Lagenodelphis hosei

F: Dauphin de Fraser
ESP: Delfín chato

BEHAVIOUR: Fraser's Dolphins generally occur in large pods of between 100 and 1,000 animals, although much smaller groups are encountered occasionally. At the surface, Fraser's Dolphins are often highly active, creating a burst of spray as they surface rapidly, often porpoising and breaching. They are gregarious and frequently occur in mixed pods with several other species. Unlike many other dolphins, Fraser's Dolphins generally feed at depths of over 200 m and so rarely associate with seabirds or schools of tuna.

Adult length:	2·0–2·6 m
Group size:	100–1,000
Behaviour:	Highly active and acrobatic

IDENTIFICATION:
Fraser's Dolphin has a unique shape, being small and stocky with a short but distinct beak. The relatively small dorsal fin is almost triangular in shape and in males can appear to be angled slightly forwards. The flippers and flukes are also relatively small.

The patterning and coloration of the body is highly characteristic, being bluish-grey above and pinkish-white below. A thick, dark stripe, often beginning as a 'mask', extends from the eye and along the flank as far as the anus. This stripe becomes wider and darker with age, being least defined in juveniles and most striking in adult males. A pale cream-coloured line runs from the forehead along the upper edge of the lateral flank stripe, and a dark stripe runs between the base of the flipper and the lower jaw.

STATUS AND DISTRIBUTION: The status and distribution of Fraser's Dolphin remains poorly understood, but it occurs throughout the world and is generally restricted to deep tropical waters. In Europe, Fraser's Dolphin has been sighted at sea only as far north as the Azores and Madeira. The only record for the European Atlantic involved a mass stranding of 11 individuals near Treguier on the north coast of Brittany, France, in May 1984. Analysis of stomach contents suggested that the animals had been feeding on fish and squid over the European continental shelf before they stranded. The only other record of this species for northern Europe involved a stranded animal discovered on South Uist, Western Isles, Scotland on 3rd September 1996.

Male Fraser's Dolphin.

SIMILAR SPECIES: At a distance, Fraser's Dolphin can be confused with other oceanic dolphins of a similar size and shape, including Short-beaked Common Dolphin (*page 50*) and Striped Dolphin (*page 52*). However, the distinctive markings of Fraser's Dolphin means that, in good light, confusion is only likely with Striped Dolphin, which is less robust, has a larger dorsal fin, a longer beak, a thinner flank stripe and a pale dorsal blaze.

Glossary

Baleen	Comb-like plates which grow from the upper jaw of some large whales. They are used to filter food from sea water.
Beak	The snout, or forward projecting jaws of a cetacean.
Blow	The visible exhalation of a whale or dolphin.
Blowhole	The nostril(s) of a cetacean.
Blubber	The layer of insulating fat below the skin of marine mammals.
Bow-riding	Swimming in the bow pressure-wave created by moving vessels or large whales.
Breach	Leaping clear or partially clear of the surface, usually landing with a splash.
Bull	An adult male whale or dolphin.
Calf	Young cetacean which is still being nursed by its mother.
Callosity	Rough, lumpy protrusion on the head of a right whale.
Cape	Darker region on the back of some cetaceans.
Cephalopod	Member of the Class Cephalopoda, including cuttlefish, squid and octopus.
Cetacean	A marine mammal of the Order Cetacea, which includes all the whales, dolphins and porpoises.
Chevron	A pale marking, shaped like a letter V, on the back of some Fin Whales.
Cow	An adult female whale.
Delphinid	Collective term for all dolphin-like odontocetes.
Dolphin	Small cetacean which usually has a beak, conical-shaped teeth, and a falcate dorsal fin.
Dorsal	A directional term meaning towards the back or upper-side.
Dorsal fin	The upper or top fin in marine vertebrates.
Echolocation	The location of objects by reflected sound.
Falcate	Sickle-shaped and curved backwards.
Flipper	The forelimbs of marine mammals, including cetaceans, pinnipeds and manatees.
Flipper-slapping	Raising the flipper or pectoral fin out of the water and slapping it on the surface.
Flukes	The horizontally flattened tail of cetaceans that function as an organ of propulsion.
Juvenile	Young cetacean that is independent of its mother, but is not fully mature.
Krill	Small, shrimp-like crustaceans.
Lob-tailing	Raising tail flukes into the air and slapping them on the surface.

Logging	The act of lying still on or near the surface.
Lunge-feeding	Baleen whales occasionally exhibit this behaviour. Their heads lunge vertically out of the water, jaws wide open in an attempt to surprise and engulf schooling prey.
Melon	The bulbous forehead of some toothed cetaceans.
Mysticeti	The Order of baleen whales.
Odontoceti	The Order of toothed whales.
Pectoral fin	The flippers or forelimbs of a cetacean.
Pelagic	Oceanic, inhabiting the open sea; living in the surface waters or middle depths of the ocean.
Pod	A discrete, coordinated group of cetaceans.
Porpoise	Small cetacean with an indistinct beak or no beak, spade-shaped teeth, and usually a small, triangular dorsal fin.
Porpoising	Leaping clear of the water while swimming rapidly forwards.
Rorqual	A baleen whale of the genus Balaenoptera.
Rostrum	The beak or snout of a baleen whale.
Sounding-dive	Deep dive, usually following a series of shorter, shallower dives.
Splash guard	Area around the blowhole of some whales which is raised during the breathing sequence.
Spout	A column of spray thrown into the air by a whale when breathing.
Spy-hopping	A behaviour that involves a cetacean raising its head vertically above the water surface.
Stranding	When a cetacean comes ashore, either dead or alive.
Tail-stock	The tapered rear part of the body from behind the dorsal fin to the tail flukes.
Temperate	Mid-latitude region, between the polar and sub-tropical waters, characterised by a mild climate.
Terminal dive	Deep dive, usually following a series of short, shallower dives, when flukes are most likely to be seen.
Throat grooves	Grooves or pleats in the throat region of some whales.
Transient	Group of cetaceans which move through an area as opposed to resident populations.
Tubercles	Raised knobs or lumps, such as those on the head of a Humpback Whale.
Ventral	Pertaining to the underside of the body.
Ventral pleats	The longitudinal grooves on the undersurface of some species of baleen whale.

Cetacean conservation – and how you can get involved

Despite being on the very doorstep of the UK, Ireland, France, and Spain, for many years the European Atlantic remained one of the world's best-kept whale-watching secrets.

However, from the early 1990s, the commissioning of two ferry routes operating across the English Channel and Bay of Biscay, and the rapid development of whale-watching in south-west England, Wales and Ireland, vastly improved our access to the ocean realm. Suddenly, we realised just how much was out there! The result has been a sea-change in our perception of these waters, which are now recognized as truly world-class, perhaps supporting the highest diversity of whales, dolphins and porpoises in the eastern North Atlantic.

Sadly, as scientists and naturalists have spent more time amongst these majestic mammals, we have also become increasingly aware of how humans have managed to degrade even the most distant ocean wildernesses. The endless stream of plastic bags, bottles and other rubbish, the fresh carcasses of dolphins floating near trawler vessels, and the serious bodily injuries to cetaceans from interactions with propellers and fishing gear, represent just a few of the most obvious signs that man is having a significant negative impact on marine ecosystems. Thankfully though, there is also room for optimism, with an ever-increasing number of people involved in watching, studying and working to protect marine mammals and their habitats across the European Atlantic (see the list on *page 83*).

One of the most exciting attributes of several cetacean conservation and research organisations working in the region is the strong focus on public participation. Nowhere has this been more prevalent than among a number of charities that have developed long-term monitoring programmes onboard ferries. These organisations encourage the public to get involved by providing survey training, followed by low-cost opportunities to see whales and dolphins at sea. The data gathered are then utilised by students and researchers in order to answer important ecological or conservation questions. Together, these organisations form a partnership called the Atlantic Research Coalition (ARC). Most marine conservation charities also rely upon volunteers to assist them with other vital work. Whether becoming involved in research or fundraising, organising events or training volunteers, you can also help to make a difference.

It is also possible to get actively involved with whales and dolphins without even leaving the shore, as live and dead cetaceans regularly wash up along most coasts in the region. There are a number of marine mammal stranding networks set up to attend these animals. On such occasions, important information, including post-mortem results, may be gained from carcasses in order to ascertain the cause of death. If the animal live-strands, its health is assessed in order to judge whether an attempted refloatation should be made. These stranding networks welcome new members and provide excellent training courses.

Cetaceans and the seas that form their home face an uncertain future at the hands of mankind. If future generations are to enjoy watching these superb mammals as we do, we must redouble our efforts to ensure a sustainable future for all marine life. The first encouraging step down this road is an improved understanding of the threats faced by wildlife at sea, the development of appropriate conservation initiatives, and the involvement of the public. Thankfully, there are an increasing number of people who are working towards, or care passionately about, marine conservation in the European Atlantic. They serve to remind us that the future of this enchanting stretch of ocean is in our own hands.

Further reading

A Field Guide to Whales, Porpoises and Seals from Cape Cod to Newfoundland. S.K. Katona, V. Rough & D.T. Richardson. 1993. Smithsonian IP, USA.

A guide to the identification of the Whales and Dolphins of Ireland. J. Wilson & S. Berrow. 2006. The Irish Whale and Dolphin Group (IWDG), Ireland.

A guide to the whales, dolphins and porpoises of the United Kingdom (2000) The first WDCS annual report on the status of UK cetaceans. D. Walker & M. De Boer. 2003. Whale and Dolphin Conservation Society (WDCS), UK.

Dolphins and Whales from the Azores. S. Vialle. 1997. Portugal.

Encyclopedia of Marine Mammals. W.F. Perrin, B. Würsig & J.M. Thewissen (Eds.). 2002. Academic Press, USA.

Eyewitness Handbook of Whales, Dolphins and Porpoises. M. Carwardine. 1995. Dorling Kindersley, London.

Guide to the Identification of Whales, Dolphins and Porpoises in European Seas. P.G.H. Evans. 1995. Scottish Natural Heritage, UK.

Whales and Dolphins: Guide to the Biology and Behaviour of Cetaceans. M. Würtz & N. Repetto. 1998. Swan Hill Press, UK.

The Best Whale Watching in Europe. A guide to seeing whales, dolphins and porpoises in all European waters. E. Hoyt. 2003. Whale and Dolphin Conservation Society (WDCS), UK.

ORCA No. 3. The Annual Report of Organisation Cetacea. D. Walker (Ed.). 2004. Organisation Cetacea (ORCA), UK.

ORCA No. 2. Incorporating a report on the whales, dolphins and seabirds of the Bay of Biscay and English Channel. G. Cresswell & D. Walker. 2002. Organisation Cetacea (ORCA), UK.

ORCA No 1. A Report on the Whales, Dolphins and Seabirds of the Bay of Biscay and the English Channel. G. Cresswell & D. Walker (Eds.). 2001. Organisation Cetacea (ORCA), UK.

Mark Carwardine's Guide to Whale Watching. Britain and Europe: Where to go, what to see. M. Carwardine. 2003. New Holland Publishers, UK.

Whales and Dolphins of Great Britain. D. Walker & A. Wilson. 2007. Cetacea Publishing.

Whales and Dolphins of the North American Pacific. G. Cresswell, D. Walker & T. Pusser. 2007. WILDGuides Ltd, UK.

Whales and Dolphins of the North Sea. K. Camphuysen, G. Peet, & F. Maas. 2007. Fontaine Uitgevers BV, Holland.

Whales and Dolphins – The Ultimate Guide to Marine Mammals. M. Carwardine, E. Hoyt, R. Ewan Fordyce & P. Gill. 1998. Harper Collins.

Whales, Dolphins and Seals. A Field Guide to the Marine Mammals of the World. H. Shirihai & B. Jarrett. 2006. A&C Black Publishers Ltd, London.

Whalewatcher – a global guide to watching whales, dolphins and porpoises in the wild. T. Day. 2006. The Natural History Museum, London.

Useful websites

A Bay to Remember
www.baytoremember.co.uk

Beaked Whale Resource
www.beakedwhaleresource.com

Breathtaking Whales
www.breathtakingwhales.com

Brittany Ferries
www.brittany-ferries.co.uk

British Divers Marine Life Rescue
www.bdmlr.org.uk

Centre for Research into Marine Mammals
http://crmm.univ-lr.fr

Cetacea Publishing
www.cetaceapublishing.com

Dolphin Survey Boat Trips
www.cbmwc.org

Durlston Country Park
www.durlston.co.uk (then go to 'marine')

Earthwatch Institute
www.earthwatch.org

European Cetacean Society
www.europeancetaceansociety.eu/ecs

Fowey Marine Adventures
www.fma.fowey.com

Gower Marine Mammals Project
www.gmmp.org.uk

Greenpeace
www.greenpeace.org

International Fund for Animal Welfare
www.ifaw.org/ifaw

International Whaling Commission
www.iwcoffice.org

Irish Whale and Dolphin Group
www.iwdg.ie

Marine Connection
www.marineconnection.org

Marine Conservation Society
www.mcsuk.org

Marine Discovery Penzance
www.marinediscovery.co.uk

Marinelife
www.biscay-dolphin.org.uk

Newquay Boat Trips
www.newquayboattrips.co.uk

Oceanopolis
www.oceanopolis.com/2008_intro.php

ORCA – Organisation Cetacea
www.orcaweb.org.uk

P&O Ferries
www.poferries.com/tourist

Planet Whale
www.planetwhale.com

Sea Mammal Research Unit
http://smub.st-andrews.ac.uk

Seaquest Southwest
www.cornwallwildlifetrust.org.uk/nature/marine
(then select 'marine life and search for Seaquest')

Seawatch Foundation
www.seawatchfoundation.org.uk

Silver Dolphin Centre
www.silverdolphinmarine
conservationanddiving.co.uk

Spanish Cetacean Society
www.cetaceos.com

The Company of Whales
www.companyofwhales.co.uk

Thousand Islands Expeditions
www.thousandislands.co.uk

UKCetnet
http://tech.groups.yahoo.com/group/ukcetnet

Ultimate Pelagics
www.ultimatepelagics.com

Venture Jet Ltd
www.venturejet.co.uk/

Whale and Dolphin Conservation Society
www.wdcs.org

Whale-Watching-Web
www.physics.helsinki.fi/whale

WILDGuides
www.wildguides.co.uk

Wise Scheme
www.wisescheme.org

World Wide Fund for Nature
www.panda.org

Photographic and artwork credits

FRONT COVER:
Short-beaked Common Dolphins:
Hugh Harrop / www.hughharrop.com.

p.1 **Killer Whale:** Dylan Walker.
p.4 **Short-beaked Common Dolphins:** Dylan Walker.
p.9 **Killer Whales:** Hugh Harrop / www.hughharrop.com.
p.11 **Sea states 0–9:** Dylan Walker.
p.12 **Humpback Whale and White-beaked Dolphin:** Swimming beside whale-watching boat (Gannet artwork superimposed by Rob Still); Dylan Walker. **Minke Whale:** Male; Dylan Walker. **Cuvier's Beaked Whale:** Male; Dylan Walker.
p.13 **Bottlenose Dolphins:** In normal and evening light, CIRCE. **Minke Whale:** Dylan Walker. **Cuvier's Beaked Whale:** Graeme Cresswell. **Short-finned Pilot Whale:** Dylan Walker. **Short-beaked Common Dolphin:** Dylan Walker. **Feeding flock of Kittiwakes and Northern Gannets:** Dylan Walker. **Fishing boat:** Dylan Walker. **Fin Whale:** Blows; Dylan Walker.
p.14 **Humpback Whale:** Graeme Cresswell. **Long-finned Pilot Whale:** Spy-hopping, CIRCE. **Long-finned Pilot Whale:** Tail-slapping; Graeme Cresswell.
p.15 **Looking forwards from P&O cruise-ferry Pride of Bilbao:** Dylan Walker.

p.16 **Ferry-based whale-watchers in the Bay of Biscay:** Dylan Walker.
p.17 **Whale-watchers at Galley Head, Ireland:** Pádraig Whooley.
p.18 **North Atlantic Right Whale:** Robin Baird.
p.19 **Fin Whale:** Hugh Harrop / www.hughharrop.com. **Sperm Whale:** Hugh Harrop / www.hughharrop.com.
p.20 **Gervais' Beaked Whale:** Sergio Hanquet. **Long-finned Pilot Whale:** Graeme Cresswell.
p.21 **Short-beaked Common Dolphin:** Dylan Walker. **Harbour Porpoise:** Graeme Cresswell.

THE PLATES

p.25 **Minke Whale:** Head and back, back showing pointed dorsal fin, blow; Graeme Cresswell. Back showing blunt dorsal fin; Dylan Walker.
p.27 **Sei Whale:** All images; Alan Henry.
p.29 **Fin Whale:** Showing right fin, dorsal fin and tail-stock, back; Hugh Harrop / www.hughharrop.com. Showing left jaw, Blow; Graeme Cresswell, Dylan Walker.
p.31 **Humpback Whale:** Breaching, uppersides of tail flukes; Graeme Cresswell. Blow, undersides of tail flukes, tail-stock, back and dorsal fin; Dylan Walker.
p.33 **Sperm Whale:** All images; Susan & David Merrett.

p.37 **Northern Bottlenose Whale**: *Head, dorsal fin*;
Susan & David Merrett. *Head and beak*; Graeme Cresswell.
Back and dorsal fin; Dylan Walker.

p.39 **Cuvier's Beaked Whale**: *All lightly scarred individuals*;
Graeme Cresswell. *Heavily scarred male with one protruding tooth*; Dave Sellwood.

p.43 **Sowerby's Beaked Whale**: *Swimming away*; Graeme Cresswell.
Beak and melon; Matt Hobbs.

p.43 **True's Beaked Whale**: *Breaching*; Dylan Walker.

p.45 **Killer Whale**: *All images*; Graeme Cresswell.

p.47 **Long-finned Pilot Whale**: *Tail-stock*; Hugh Harrop / www.hughharrop.com. *Pair swimming*; Graeme Cresswell.

p.51 **Short-beaked Common Dolphin**: *Porpoising*; Hugh Harrop / www.hughharrop.com. *From above, upper body*; Dylan Walker.

p.53 **Striped Dolphin**: *All images*; Hugh Harrop / www.hughharrop.com.

p.55 **Bottlenose Dolphin**: *Porpoising*; Hugh Harrop / www.hughharrop.com. *Head and back*; Dave Sellwood. *Back and dorsal fin*; Dylan Walker.

p.57 **Risso's Dolphin**: *Grey immature*; Susan and David Merrett. *Head and back, back and dorsal fin*; Graeme Cresswell.

p.59 **Atlantic White-sided Dolphin**: *Head, back*; Hugh Harrop / www.hughharrop.com. *Two swimming*; Sascha Hooker.

p.61 **White-beaked Dolphin**: *From above*; Graeme Cresswell. *Dorsal fin, back*; Dylan Walker.

p.63 **Harbour Porpoise**; *All images*; Graeme Cresswell.

p.65 **Beluga**: Graeme Cresswell.

p.67 **Blue Whale**: *Rostrum*; Dave Sellwood. *Tail-stock, tail flukes*; Ian Rowlands. *Blow, back and dorsal fin*; Graeme Cresswell.

p.69 **North Atlantic Right Whale**: *Tail flukes*; Robin Baird. *Blow*; Kristi M. Willis. *Head*; Brennan Mulrooney. *Head and back*; dotbleu.

p.70 **Pygmy Sperm Whale**: Robert L. Pitman.

p.71 **Dwarf Sperm Whale**: C.D. Macleod.

p.72 **Gervais' Beaked Whale**: *All images*; Sergio Hanquet.

p.73 **Blainville's Beaked Whale**: *All images*; Graeme Cresswell.

p.74 **Short-finned Pilot Whale**: *Tail stock, blow and back*; Dylan Walker. *Back and dorsal fin*; Graeme Cresswell.

p.75 **False Killer Whale**: *Back and dorsal fin*; Susan & David Merrett. *Head*; Jeremy Kiszka.

p.76 **Pygmy Killer Whale**: *Head, dorsal fin*; Wild Dolphin Foundation.

p.77 **Melon-headed Whale**: *All images*; Jeremy Kiszka.

p.79 **Rough-toothed Dolphin**: *Group*; Graeme Cresswell. *Breaching*; Dylan Walker.

p.79 **Fraser's Dolphin**: *Male*; Jeremy Kiszka.

Megaptera novaeangliae	30
Melon-headed Whale	35, **77**, 76, 77
Mesoplodon	40
Mesoplodon bidens	**42**
——— *densirostris*	**73**
——— *europaeus*	**72**
——— *mirus*	**43**
Minke Whale	10, *12*, *13*, 22, **24**, 25, 27, 29
North Atlantic Right Whale	*18*, 23, **68**
Northern Bottlenose Whale	10, 34, 25, **36**
Orca	23, 35, **44**
Orcinus orca	**44**
Peponocephala electra	**77**
Phocoena phocoena	**62**
Phocoenidae	21
Physeter macrocephalus	**32**
Physeteridae	19
Porpoise, Harbour	10, *21*, 23, 48, **62**
Pseudorca crassidens	**75**
Pygmy Killer Whale	35, **76**, 77
Pygmy Sperm Whale	49, **70**, 71
Risso's Dolphin	10, 49, **56**
Rough-toothed Dolphin	48, **78**
Sei Whale	10, 22, 25, **26**, 29
Short-beaked Common Dolphin	10, *13*, *21*, 23, 48, **50**
Short-finned Pilot Whale	35, 47, **74**
Sowerby's Beaked Whale	10, 34, 41, **42**, 72
Sperm Whale	10, *19*, 23, **32**
Stenella coeruleoalba	**52**
Steno bredanensis	**78**
Striped Dolphin	10, 48, **52**
True's Beaked Whale	34, 41, **43**, 72
Tursiops truncatus	**54**
Whale, Blainville's Beaked	34, 41, **73**
———, Blue	22, **66**
———, Cuvier's Beaked	10, *12*, *13*, 34, **38**
———, Dwarf Sperm	49, **71**
———, False Killer	35, 47, **75**
———, Fin	10, *13*, *19*, 22, 25, 27, **28**, 29
———, Gervais' Beaked	20, 34, 41, **72**
———, Humpback	*12*, *14*, 23, **30**
———, Killer	10, 9, 23, 35, **44**
———, Long-finned Pilot	10, *13*, *14*, 20, 35, **46**, 47
———, Melon-headed	35, **77**
———, Minke	10, *12*, *13*, 22, **24**, 25, 27, 29
———, North Atlantic Right	*18*, 23, **68**
———, Northern Bottlenose	10, 34, 25, **36**
———, Pygmy Killer	35, **76**, 77
———, Pygmy Sperm	49, **70**, 71
———, Sei	10, 22, 25, **26**, 29
———, Short-finned Pilot	35, 47, **74**
———, Sowerby's Beaked	10, 34, 41, **42**, 72
———, Sperm	10, *19*, 23, **32**
———, True's Beaked	34, 41, **43**, 72
White-beaked Dolphin	10, *12*, 48, **60**
Ziphiidae	20
Ziphius cavirostris	**38**